EXAMKRACKERS
1001 Questions
in
MCAT
Organic Chemistry

OSOTE
PUBLISHING

ISBN 1-893858-19-7

2006 Edition

examkrackers.com
osote.com
audioosmosis.com

Printed and bound in the U.S.A.

Table of Contents

PHYSICAL SCIENCES

DIRECTIONS. Most questions in the Physical Sciences test are organized into groups, each preceded by a descriptive passage. After studying the passage, select the one best answer to each question in the group. Some questions are not based on a descriptive passage and are also independent of each other. You must also select the one best answer to these questions. If you are not certain of an answer, eliminate the alternatives that you know to be incorrect and then select an answer from the remaining alternatives. Indicate your selection by blackening the corresponding oval on your answer document. A periodic table is provided for your use. You may consult it whenever you wish.

PERIODIC TABLE OF THE ELEMENTS

1 **H** 1.0																	2 **He** 4.0
3 **Li** 6.9	4 **Be** 9.0											5 **B** 10.8	6 **C** 12.0	7 **N** 14.0	8 **O** 16.0	9 **F** 19.0	10 **Ne** 20.2
11 **Na** 23.0	12 **Mg** 24.3											13 **Al** 27.0	14 **Si** 28.1	15 **P** 31.0	16 **S** 32.1	17 **Cl** 35.5	18 **Ar** 39.9
19 **K** 39.1	20 **Ca** 40.1	21 **Sc** 45.0	22 **Ti** 47.9	23 **V** 50.9	24 **Cr** 52.0	25 **Mn** 54.9	26 **Fe** 55.8	27 **Co** 58.9	28 **Ni** 58.7	29 **Cu** 63.5	30 **Zn** 65.4	31 **Ga** 69.7	32 **Ge** 72.6	33 **As** 74.9	34 **Se** 79.0	35 **Br** 79.9	36 **Kr** 83.8
37 **Rb** 85.5	38 **Sr** 87.6	39 **Y** 88.9	40 **Zr** 91.2	41 **Nb** 92.9	42 **Mo** 95.9	43 **Tc** (98)	44 **Ru** 101.1	45 **Rh** 102.9	46 **Pd** 106.4	47 **Ag** 107.9	48 **Cd** 112.4	49 **In** 114.8	50 **Sn** 118.7	51 **Sb** 121.8	52 **Te** 127.6	53 **I** 126.9	54 **Xe** 131.3
55 **Cs** 132.9	56 **Ba** 137.3	57 **La*** 138.9	72 **Hf** 178.5	73 **Ta** 180.9	74 **W** 183.9	75 **Re** 186.2	76 **Os** 190.2	77 **Ir** 192.2	78 **Pt** 195.1	79 **Au** 197.0	80 **Hg** 200.6	81 **Tl** 204.4	82 **Pb** 207.2	83 **Bi** 209.0	84 **Po** (209)	85 **At** (210)	86 **Rn** (222)
87 **Fr** (223)	88 **Ra** 226.0	89 **Ac**† 227.0	104 **Unq** (261)	105 **Unp** (262)	106 **Unh** (263)	107 **Uns** (262)	108 **Uno** (265)	109 **Une** (267)									

	58 **Ce** 140.1	59 **Pr** 140.9	60 **Nd** 144.2	61 **Pm** (145)	62 **Sm** 150.4	63 **Eu** 152.0	64 **Gd** 157.3	65 **Tb** 158.9	66 **Dy** 162.5	67 **Ho** 164.9	68 **Er** 167.3	69 **Tm** 168.9	70 **Yb** 173.0	71 **Lu** 175.0
†	90 **Th** 232.0	91 **Pa** (231)	92 **U** 238.0	93 **Np** (237)	94 **Pu** (244)	95 **Am** (243)	96 **Cm** (247)	97 **Bk** (247)	98 **Cf** (251)	99 **Es** (252)	100 **Fm** (257)	101 **Md** (258)	102 **No** (259)	103 **Lr** (260)

Structural Formula

1. What is the correct Lewis dot structure for CH_3CH_2CHO?

   A.

   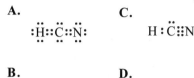

   C.

   H H Ö:H
   H:C:C:C:
   H H

   B.

   H H ·Ö·
   H:C:C::C:H
   H H

   D.

   H H
   H:C:C::C::Ö:H
   H H

2. What is the correct Lewis dot structure for HCN?

   A.

   :H::C::N:

   C.

   H:C̈::N

   B.

   H:C̈::N̈

   D.

   H:C::N̈

3. A Lewis base donates a pair of electrons. Which of the following molecules could be a Lewis base?

   A. CH_4
   B. NH_3
   C. $AlCl_3$
   D. BH_3

4. Which of the following molecules could NOT be a Lewis base?

   A. $NH(CH_3)_2$
   B. NH_4^+
   C. H_2O
   D. OH^-

Refer to the Newman projection below to answer questions 5-7.

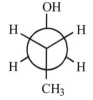

5. In the Newman projection, what does the circle represent?

   A. The largest carbon in the structure.
   B. The second carbon along the axis of the bond.
   C. The first carbon along the axis of the bond.
   D. The radius of the bond.

6. In the Newman projection, what does the intersection of the three lines represent?

   A. The largest carbon in the structure.
   B. The second carbon along the axis of the bond.
   C. The first carbon along the axis of the bond.
   D. The radius of the bond.

7. In the Newman projection, the hydroxide group and the methyl group are:

   A. attached to the same carbon.
   B. attached to different carbons.
   C. attached to carbons with no hydrogens attached.
   D. attached directly to each other.

Refer to the Fischer projection below to answer questions 8-10.

8. In the Fischer projection, the hydroxide groups are:

   A. coming out of the page on the right side and into the page on the left.
   B. coming out of the page on the left side and into the page on the right.
   C. coming out of the page.
   D. going into the page.

9. In the Fischer projection, the methyl groups are:

   A. coming out of the page on the top and into the page on the bottom.
   B. coming out of the page on the top and into the page on the bottom.
   C. coming out of the page.
   D. going into the page.

10. In the Fischer projection, what do the intersections of the vertical and horizontal lines represent?

    A. rotation of the bond
    B. carbon atoms
    C. double bonds
    D. nothing

Refer to the dash-line-wedge formula below to answer questions 11-13.

11. In the dash-line wedge formula what atom(s) is (are) coming out of the page?

   I. Cl
   II. Br
   III. H

   A. I only
   B. II only
   C. II and III
   D. I and III

12. In the dash-line wedge formula what atom(s) is (are) going back into the page?

   I. Cl
   II. CH_3
   III. H

   A. I only
   B. II only
   C. II and III
   D. I and III

13. In the dash-line wedge formula what atom(s) is (are) in the plane of the page?

   I. CH_3
   II. Br
   III. H

   A. I only
   B. II only
   C. II and III
   D. I and III

14. Which of the following structures is an amide?

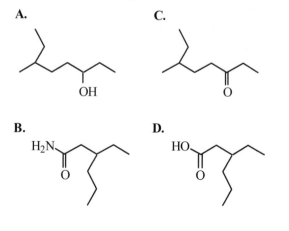

15. Which of the following structures is a ketone?

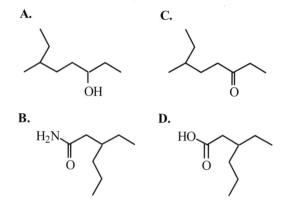

16. What is the name of the functional group contained in the structure shown below?

   A. ether
   B. ester
   C. carboxylic acid
   D. alcohol

17. What is the name of the functional group contained in the structure shown below?

   A. ether
   B. ester
   C. carboxylic acid
   D. alcohol

18. Isoamylacetate is a honeybee pheromone that is released on the skin when a bee stings its victim. This pheromone has a sweet smell and attracts other bees to join the fight. The structure of isoamylacetate is shown below. What functional group does it contain?

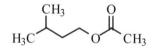

   A. ether
   B. ester
   C. carboxylic acid
   D. alcohol

19. Tetrahydrofuran (THF) is a common solvent for synthetic reactions because it is an ether. Ethers are relatively inert so they do not participate in the reaction. Which of the following structures is THF?

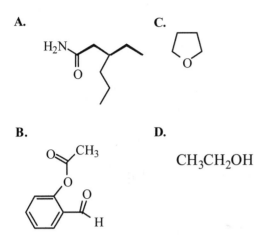

A.

C.

B.

D.

CH_3CH_2OH

20. Aspirin is the common name for acetylsalicylic acid. The structure is shown below, what three functional groups are present in this molecule?

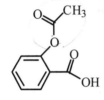

A. ether, aromatic ring, carboxylic acid
B. ester, phenyl, carboxylic acid
C. alkene, aromatic ring, carbooxylic acid
D. alcohol, ether, carboxylic acid

21. Luminol is a compound used by the police to detect the presence of dried blood. Upon oxidation, luminol emits a green light in the presence of a metal ion such as the Fe ion found in blood. The structure of luminol is shown below. How many amine groups are present in this molecule?

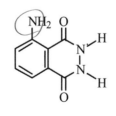

A. 0
B. 1
C. 2
D. 3

22. How many amide groups are present in luminol? See question 21 for the structure?

A. 0
B. 1
C. 2
D. 3

23. Menthol is one of the molecules responsible for the mint odor. What functional groups are present in menthol?

A. cycloalkane and alkene
B. cycloalkene and alcohol
C. cycloalkane and ether
D. cycloalkane and alcohol

24. Similar in structure to menthol is limonene which is responsible for the lemon odor found in natural products. What is the difference in functional groups between these two structures?

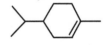

A. Menthol has an alcohol and limonene has an alkyne.
B. Menthol has an alcohol and limonene has an alkene.
C. Menthol has an ether and limonene has an alkyne.
D. Menthol has an ether and limonene has an alkene.

25. Eugenol is a dental anesthetic that is isolated from oil of cloves. The structure is shown below. What functional groups are present in this toothache remedy?

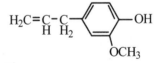

A. aromatic ring, alkene, alcohol, and ether
B. cycloalkyl ring, alkene, alcohol, and ester
C. aromatic ring, alkyne, ether, and ester
D. aromatic ring, alkene, alcohol, and ester

26. When eugenol (shown in question 25) is isolated from the clove flower buds, acetyleugenol (shown below) is also isolated and must be separated from the eugenol. What is the difference between these two compounds?

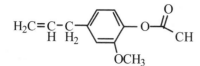

   A. The ether of eugenol is converted to an ester in acetyleugenol.
   B. The alcohol of eugenol is converted to an ether in acetyleugenol.
   C. The ester of eugenol is converted to an ether in acetyleugenol.
   D. The alcohol of eugenol is converted to an ester in acetyleugenol.

27. What is the index of hydrogen deficiency for the molecular formula C_8H_{12}?

   A. 0
   B. 1
   C. 2
   D. 3

28. What is the index of hydrogen deficiency for the molecular formula $C_8O_2H_{12}$?

   A. 0
   B. 1
   C. 2
   D. 3

29. What is the index of hydrogen deficiency for the molecular formula $C_8O_2H_{11}Br$?

   A. 0
   B. 1
   C. 2
   D. 3

30. What is the index of hydrogen deficiency for the structure shown below?

   A. 1
   B. 2
   C. 3
   D. 4

31. What is the index of hydrogen deficiency for the structure shown below?

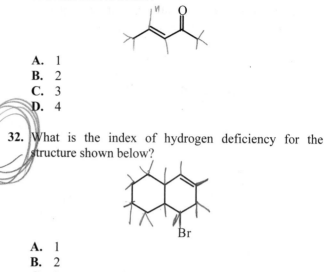

   A. 1
   B. 2
   C. 3
   D. 4

32. What is the index of hydrogen deficiency for the structure shown below?

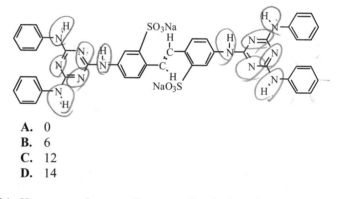

   A. 1
   B. 2
   C. 3
   D. 4

33. A brightener that is used for white shirts is Calcofluor White MR. The structure of this whitener is shown below. How many amine functionalities are present in this molecule?

   A. 0
   B. 6
   C. 12
   D. 14

34. How many degrees of unsaturation (index of hydrogen deficiency) are in each phenyl ring on the end of the calcofluor shown in question 33?

   A. 1
   B. 2
   C. 3
   D. 4

35. Another brightener that is used on wool and nylon fabrics is 7-dimehtylamino-4-methylcoumarin. What functional groups are present in this brightener?

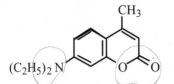

A. aromatic ring, alkene, ketone, ether and amide
B. aromatic ring, cycloalkene, ketone, and ether
C. aromatic ring, cycloalkene, ester, and amine
D. aromatic ring, cycloalkene, ester, and amide

36. What is the IUPAC name for the structure shown below?

A. 2,2-dimethyl-3-n-butylpentane
B. 4,4-dimethyl-3-n-butylpentane
C. 5-tert-butylheptane
D. 3-tert-butylheptane

37. What is the IUPAC name for the structure shown below?

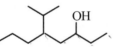

A. 4-isopropyl-6-octanol
B. 5-isopropyl-3-octanol
C. 5-isopropyl-3-hexanol
D. 4-2-butanol-5-methylhexane

38. Which structure below is the correct structure of 1-ethyl-3-(1,1,3-trimethylbutyl)cyclooctane?

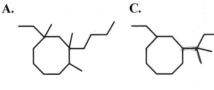

39. Ethanol reacts with sulfuric acid at 140°C to form an ether. The ether formed would be:

A. diethylether
B. 1-ethyne
C. 1-propene
D. ethanal

40. Alcohols are converted to esters using carboxylic acids. Using anhydrides is a better way to achieve this conversion. Which reagent would best convert octanol to octyl acetate, an ester that has a fruity odor?

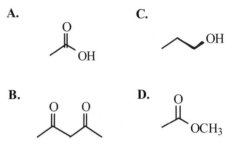

41. Bilirubin is a bile pigment found in human gallstones. The structure of bilirubin is shown below. How many amide functional groups are present in this pigment?

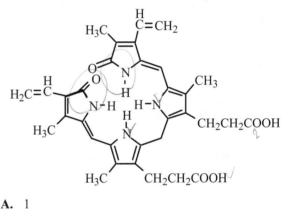

A. 1
B. 2
C. 3
D. 4

42. How many amine groups are found in bilirubin (structure shown in question 41)?

A. 1
B. 2
C. 3
D. 4

43. How many carboxylic acid groups are found in bilirubin (structure shown in question 41)?

A. 1
B. 2
C. 3
D. 4

44. 2,4-dinitrohydrazine reacts with ketones and aldehydes to form a 2,4-dinitrohyrazone. The 2,4-DNP test is used to determine if an unknown compound is an aldehyde or ketone. Which reagent would produce a positive 2,4-DNP test?

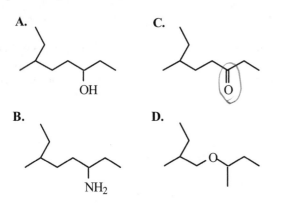

A.

C.

B.

D.

45. Given the information in question 44, which of the following compounds would be the best solvent for the 2,4-DNP test?

I.   acetone
II. ethanol
III. heptanal

A. I only
B. II only
C. II and III only
D. I and III only

46. If 1-methylcyclopentene undergoes hydration to form an alcohol, what would be the product?

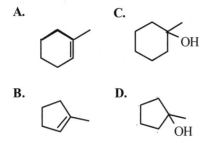

A.

C.

B.

D.

47. When 1-chloro-3-methylpentane is reacted with excess ammonia the halogen is replaced by an amine group. What is the structure of the product of this reaction?

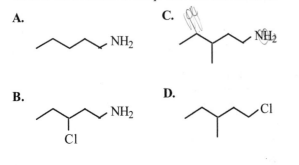

A.

C.

B.

D.

48. In the reaction shown below, what is the functional group conversion that takes place?

A. alkane to aldehyde
B. cycloalkane to cycloketone
C. alkane to cycloalkane
D. cycloalkane to cycloether

49. The following Newman projection looks down which bond of 1-choropentane, according to IUPAC carbon numbers?

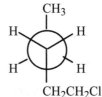

A. the bond between C1 and C2
B. the bond between C2 and C3
C. the bond between C3 and C4
D. the bond between C4 and C5

50. Which structure is the Fischer projection for the structure shown below?

A.   C.

CHO   CHO
H——OH   HO——H
CH₃   CH₃

B.   D.

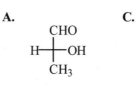   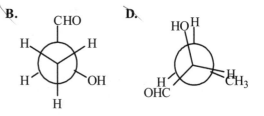

51. Which structure is the dash-line-wedge formula for the structure shown below?

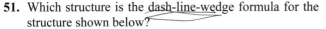

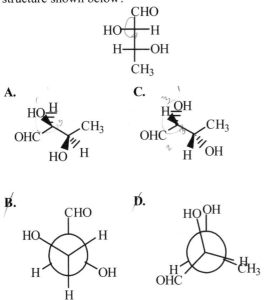

Bonding

52. What force is responsible for the bonds between atoms?

    A. Dipole-dipole
    B. Hydrogen bonding
    C. Gravitational
    D. Electrostatic

Refer to the Lewis Dot structure shown below to answer questions 53-57.

$$H:C \vdots\vdots \ddot{N}$$

53. How many electrons are involved in the sigma bond between the carbon atom and the nitrogen atom?

    A. 0
    B. 1
    C. 2
    D. 6

54. What types of bonds exist between the C atom and the N atom in HCN?

    A. 1 pi bond and two sigma bonds
    B. 2 pi bonds and one sigma bond
    C. 3 pi bonds and zero sigma bonds
    D. 3 sigma bonds and zero pi bonds

55. How many electrons does the N atom contribute to the pi bonds?

    A. 1
    B. 2
    C. 3
    D. 5

56. What is the hybridization of the sigma bond between the carbon atom and the nitrogen atom in HCN?

    A. sp
    B. sp^2
    C. sp^3
    D. sp^3d

57. What is the formal charge on the nitrogen atom in HCN?

    A. 0
    B. 1
    C. 2
    D. 3

58. Which bond is the most stable bond?

    A. sigma bond
    B. pi bond in an alkene
    C. pi bond in an alkyne
    D. pi bond in a carbonyl

59. What types of bonds are found in an alkyne functional group?

    A. one sigma bond and one pi bond
    B. two sigma bond and one pi bond
    C. one sigma bond and two pi bonds that are perpendicular to each other
    D. one sigma bond and two pi bonds that are parallel to each other

Average bond energies are shown in the table below. Use the table to answer questions 60-61.

Bond	Average Bond Energy kcal mol^{-1}
C — C	83
C = C	146
C ≡ C	200

60. Based on the values in the table, what is the approximate average bond energy for each sigma bond?

    A. 60 kcal mol^{-1}
    B. 80 kcal mol^{-1}
    C. 150 kcal mol^{-1}
    D. 200 kcal mol^{-1}

61. Based on the values in the table, what is the approximate average bond energy for each pi bond?

    A. 60 kcal mol^{-1}
    B. 80 kcal mol^{-1}
    C. 150 kcal mol^{-1}
    D. 200 kcal mol^{-1}

62. The energy of a 2sp^3 hybridized orbital for a carbon atom is

  A. higher in energy than the 2s atomic orbital and lower in energy than the 2p atomic orbital.
  B. higher in energy than the 2p atomic orbital and lower in energy than the 2s atomic orbital.
  C. higher in energy than both the 2s and the 2p atomic orbitals.
  D. lower in energy than both the 2s and the 2p atomic orbitals.

63. The overlap of what two orbitals form the pi bond between carbon atoms in an alkene?

  A. two p orbitals
  B. two sp^2 orbitals
  C. two sp^3 orbtials
  D. two s orbitals

64. How much s character does an sp^3 hybridized orbital have?

  A. no s character
  B. 25%
  C. 33.3%
  D. 50%

Refer to the structure of aspartame, an artificial sweetener, to answer questions 65-70.

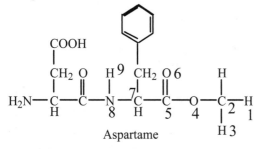

Aspartame

65. The hybridization of the C2 carbon in aspartame is:

  A. sp
  B. sp^2
  C. sp^3
  D. sp^{3}d

66. The hybridization of the C5 carbon in aspartame is:

  A. sp
  B. sp^2
  C. sp^3
  D. sp^{3}d

67. The hybridization of the N8 nitrogen in aspartame is:

  A. sp
  B. sp^2
  C. sp^3
  D. sp^{3}d

68. The bond angle formed by H1, C2, and H3 in aspartame is:

  A. 180°
  B. 120°
  C. 109°
  D. 90°

69. The bond angle formed by O4, C5, and O6 in aspartame is:

  A. 180°
  B. 120°
  C. 109°
  D. 90°

70. The bond angle formed by C7, N8, and H9 in aspartame is:

  A. 180°
  B. 120°
  C. 109°
  D. 90°

Cocaine is a stimulant that is isolated from coca leaves. Refer to the structure of cocaine shown below to answer questions 71-80.

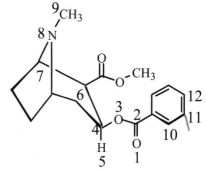

71. The hybridization of the C9 carbon in cocaine is:

  A. sp
  B. sp^2
  C. sp^3
  D. sp^{3}d

72. The hybridization of the N8 nitrogen in cocaine is:

  A. sp
  B. sp^2
  C. sp^3
  D. sp^{3}d

73. The hybridization of the O3 oxygen in cocaine is:

  A. sp
  B. sp^2
  C. sp^3
  D. sp^{3}d

74. The hybridization of the C11 oxygen in cocaine is:

    A. sp
    B. sp^2
    C. sp^3
    D. sp^{3}d

75. The bond angle formed by C10, C11, and C12 in cocaine is:

    A. 180°
    B. 120°
    C. 109°
    D. 90°

76. The bond angle formed by C2, O3, and C4 in cocaine is:

    A. 180°
    B. 120°
    C. 109°
    D. 90°

77. The bond angle with C9 in the cocaine structure at the center of the angle is:

    A. 180°
    B. 120°
    C. 109°
    D. 90°

78. How much s character is in the hybridized orbital on the C2 carbon in the cocaine structure?

    A. no s character
    B. 25%
    C. 33.3%
    D. 50%

79. How much p character is in the hybridized orbital on the N8 nitrogen in the cocaine structure?

    A. 33.3%
    B. 50%
    C. 66.6%
    D. 75%

80. What type of bond is formed between the C2 carbon atom and the O1 oxygen atom in cocaine?

    A. ionic bond
    B. pure covalent bond
    C. polar covalent bond
    D. coordinate covalent bond

81. How much p character is in the hybridized orbital on the center carbon in acetone (CH$_3$COCH$_3$)?

    A. 33.3%
    B. 50%
    C. 66.6%
    D. 75%

82. Which of the following structures have delocalized electrons?

    I.   H$_3$C–C=C–CH$_3$
              H H

    II.  ⬡

    III. H$_2$C=C–C̈–CH$_3$
              H   O

    A. I only
    B. II only
    C. I and III only
    D. II and III only

83. Which of the following structures have delocalized electrons?

    I.   H$_3$C–C=C–CH$_3$
              H H

    II.  ⬡

    III. H$_2$C=C–C–C=CH$_2$
              H H$_2$ H

    A. I only
    B. II only
    C. I and III only
    D. None of the structures.

84. What is the length of the bond between C1 and C2 in the structure shown below?

    A. the length of a carbon-carbon bond in an alkane
    B. the length of a carbon-carbon double bond in an alkene
    C. shorter than the length of a carbon-carbon double bond in an alkene
    D. between the length of a carbon-carbon bond in an alkane and the length of a carbon-carbon double bond in an alkene

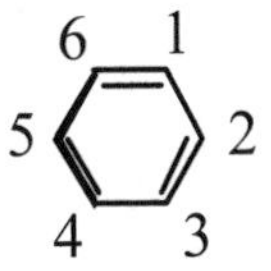

85. What is the length of the bond between C2 and C3 in the structure shown below?

A. the length of a carbon-carbon bond in an alkane
B. the length of a carbon-carbon double bond in an alkene
C. shorter than the length of a carbon-carbon double bond in an alkene
D. between the length of a carbon-carbon bond in an alkane and the length of a carbon-carbon double bond in an alkene

86. Which of the structures is NOT a resonance structure of benzene?

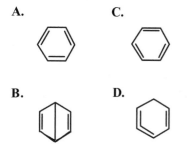

87. Which structure has the correct bond dipole found in ethanol?

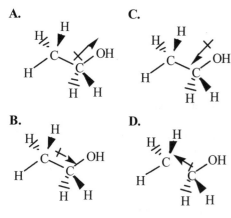

88. Which of the three molecules pictured below contain a polar bond?

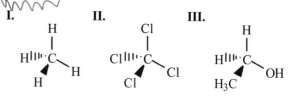

A. II only
B. III only
C. I and III only
D. I, II, and III

89. Which of the three molecules pictured below contain a net dipole moment?

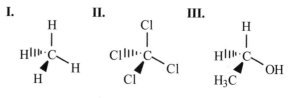

A. II only
B. III only
C. I and III only
D. II and III only

90. What is the correct net dipole moment for the structure shown below?

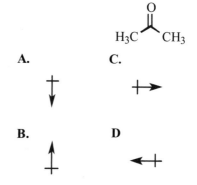

A.

B.

C.

D

91. Given the information about the dipoles of given bonds shown below, which atom has the largest difference in electronegativity compared to carbon?

Bond	Dipole Moment in Debye
C – H	0.3 D
C – N	0.22 D
C – O	0.86 D
C – Cl	1.56 D

A. H
B. N
C. O
D. Cl

92. The dipole moment for the C-Br bond is 1.5 D. What is the dipole moment for carbon tetrabromide?

    A. 0
    B. 3
    C. 6
    D. Cannot be determined.

93. What determines the polarity of a covalent bond?

    A. difference in atomic size
    B. difference in electronegativity
    C. difference in total number of protons
    D. difference in total number of valence electrons

94. The primary interaction between molecules of cyclohexane in the liquid phase is:

    A. London dispersion forces.
    B. hydrogen bonding.
    C. covalent bonding.
    D. dipole-dipole.

95. The primary interaction between molecules of <u>acetone</u> in the liquid phase is

    A. London dispersion forces.
    B. hydrogen bonding.
    C. covalent bonding.
    D. dipole-dipole.

96. The primary interaction between molecules of diethylamine in the liquid phase is:

    A. London dispersion forces.
    B. hydrogen bonding.
    C. covalent bonding.
    D. dipole-dipole.

97. Hexanol has a boiling point of 156.5°C, while hexane has a boiling point of 69°C. The reason for the dramatic difference is the boiling points is:

    A. London dispersion forces.
    B. hydrogen bonding.
    C. the larger molecular size of hexanol.
    D. the smaller molecular size of hexanol.

98. Given the dipoles shown for *trans*-1,2-dichloroethane and *cis*-1,2-dichloroethane, why is the boiling point for the trans isomer 48°C while the boiling point for the cis isomer is 60°C?

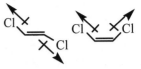

    A. The trans isomer has a greater net dipole moment leading to stronger dipole-dipole intermolecular interactions.
    B. The cis isomer has a greater net dipole moment leading to stronger dipole-dipole intermolecular interactions.
    C. The trans isomer has a greater molecular weight than the cis isomer leading to stronger London dispersion forces.
    D. The cis isomer has a greater molecular weight than the trans isomer leading to stronger London dispersion forces.

99. Which bond requires the most energy to break?

    A. sigma bond
    B. pi bond
    C. hydrogen bond
    D. dipole-dipole bond

100. Which bond requires the most energy to break?

    A. single bond
    B. double bond
    C. triple bond
    D. hydrogen bond

101. Which bond requires the most energy to break?

    A. London dispersion force bond
    B. hydrogen bonding
    C. dipole-dipole
    D. all require the same energy

102. In order to be IR active (have an absorption band in an IR spectrum) a molecule must have a dipole moment. Which of the following molecules are IR active?

   I. CH_3CH_2OH

   II.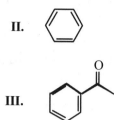

   III.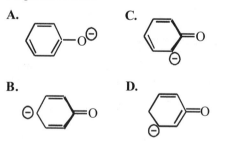

   A. II only
   B. III only
   C. I and III only
   D. II and III only

103. Conjugation in a molecule will shift the IR absorption band to a lower energy. Which of the following structures will have IR bands at a lower energy than a non-conjugated double bond?

   I. $H_2C{=}C{-}C{-}C{-}CH_3$
             $H\ H_2\ H_2$

   II. $H_2C{=}C{-}C{-}C{=}CH_2$
              $H\ H_2\ H$

   III. $H_2C{=}C{-}C{=}C{-}CH_3$
              $H\ H\ H$

   A. II only
   B. III only
   C. I and III only
   D. II and III only

104. Which of the following is <u>not</u> a resonance structure of the phenol anion?

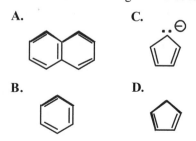

105. Which of the following structures is aromatic?

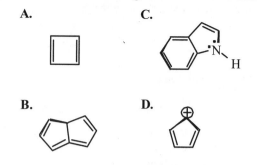

106. Which of the following structures is a <u>major</u> resonance contributor for the structure shown below?

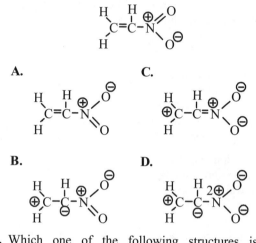

107. Which one of the following structures is <u>not</u> a resonance structure of the cation shown below?

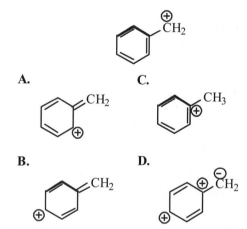

108. Which of the following structures is <u>not</u> aromatic?

   A.   **C.**

   B.   **D.**

109. An anion is stabilized by resonance and the molecule with the most stable anion after the proton is removed has the most acidic proton. Given the above statement, which of the following molecules has the most acidic proton?

A.

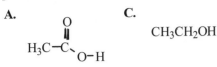

C. CH_3CH_2OH

B. CH_3CH_3

D.

$H_2C{=}\underset{H}{C}{-}CH_3$

110. In formamide the nitrogen has a bond angle of 120°, indicating sp^2 hybridization rather than the expected sp^3 hybridization. The best explanation for this observation is

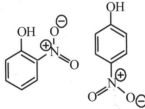

formamide

A. in one of the resonance structures the nitrogen has a double bond.
B. nitrogen always has sp^2 hybridization.
C. the lone pair of electrons on the nitrogen indicates sp^2 hybridization.
D. the nitrogen has three atoms attached therefore it must have sp^2 hybridization.

111. The structure on the left below (2-nitrophenol) is more volatile than 4-nitrophenol (the structure on the right). What is the best explanation for the difference in volatility?

A. The oxygen in the nitro group of 2-nitrophenol can hydrogen bond with the OH group of the same molecule decreasing the number of intermolecular hydrogen bonds.
B. The oxygen in the nitro group of 4-nitrophenol can hydrogen bond with the OH group of the same molecule decreasing the number of intermolecular hydrogen bonds.
C. The molecular weight of 4-nitrophenol is greater than the molecular weight of 2-nitrophenol.
D. The molecular weight of 2-nitrophenol is greater than the molecular weight of 4-nitrophenol.

Refer to the structure of Chlorophyll-a, a plant pigment, to answer questions 112 - 114.

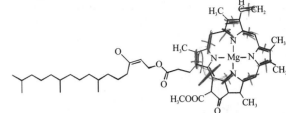

112. How many double bonds are contained in the conjugated system found in Chlorophyll-a?

A. 10
B. 11
C. 12
D. 16

113. In Chlorophyll-a, what type of bonding occurs between the Mg metal and the nitrogens of the porphorin ring?

A. covalent bond
B. ionic bond
C. coordinate covalent bond
D. hydrogen bond

114. How many sp hybridized carbons are present in Chlorophyll-a?

A. 0
B. 5
C. 15
D. 20

115. Some substitution reactions go through a trigonal planar carbocation intermediate. The center carbon has bond angles of 120°. What is the hybridization of this center carbon?

A. sp
B. sp^2
C. sp^3
D. sp^3d

116. The hydrogens on cyclopentadiene are more acidic than other alkenes. The reason for this increase in acidity is:

A. the formation of the anion creates an aromatic system.
B. the formation of the anion creates a nonaromatic system.
C. cyclic alkenes are always more acidic than straight chain alkenes.
D. the negative charge formed is not resonance stabilized.

117. The conversion of NAD$^+$ to NADH involves the reaction shown below. This conversion is an essential reaction in the chemistry of the body. What is the hybridization of the nitrogen atom in NAD$^+$ and NADH respectively?

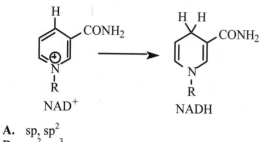

A. sp, sp^2
B. sp^2, sp^3
C. sp^3, sp^2
D. sp^3, sp

118. The fluidity of a lipid is reduced when degrees of unsaturation are added. Unsaturated lipid chains are shorter than the saturated lipid chain. What bonds determine the fluidity of the lipids?

A. dipole-dipole
B. London dispersion
C. hydrogen bonding
D. pi interactions

Stereochemistry

119. Which of the following molecules is chiral?

A. CH$_2$ClBr
B. CH$_3$CH$_2$OHCl
C. CH$_3$CH(CH$_2$CH$_3$)Cl
D. CH$_3$CH(CH$_3$)Cl

120. How are the pair of molecules shown below related?

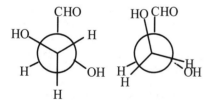

A. conformers
B. enantiomers
C. structural isomers
D. geometric isomers

121. How are the pair of molecules shown below related?

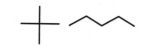

A. conformers
B. enantiomers
C. structural isomers
D. geometric isomers

122. How are the pair of molecules shown below related?

A. conformers
B. enantiomers
C. structural isomers
D. geometric isomers

123. How are the pair of molecules shown below related?

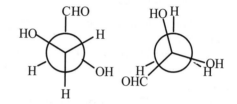

A. conformers
B. enantiomers
C. structural isomers √
D. geometric isomers

124. How are the pair of molecules shown below related?

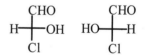

A. conformers
B. enantiomers
C. structural isomers
D. geometric isomers

125. How are the pair of molecules shown below related?

A. conformers
B. enantiomers
C. structural isomers
D. geometric isomers

126. How are the pair of molecules shown below related?

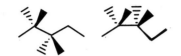

    A. conformers
    B. enantiomers
    C. structural isomers
    D. geometric isomers

127. Which of the following conformers has the highest energy?

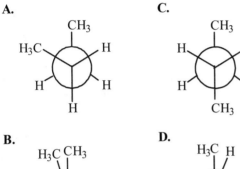

128. Which of the following conformers has the lowest energy?

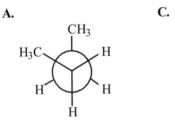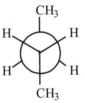

129. How many carbon-carbon bonds in the structure shown below have conformers?

    A. 1
    B. 2
    C. 3
    D. 4

130. What is the conformer shown below called?

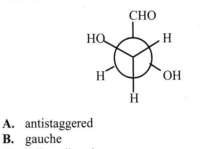

    A. antistaggered
    B. gauche
    C. fully eclipsed
    D. eclipsed

131. What is the conformer shown below called?

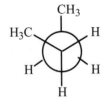

    A. antistaggered
    B. gauche
    C. fully eclipsed
    D. eclipsed

132. What is the conformer shown below called?

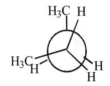

    A. antistaggered
    B. gauche
    C. fully eclipsed
    D. eclipsed

133. Geranial and neral are two terpenes that can be isolated from lemongrass. They are in part responsible for the odor. What is the relationship of these two molecules?

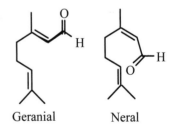

    Geranial   Neral

    A. conformers
    B. enantiomers
    C. structural isomers
    D. geometric isomers

134. Which of the following molecules is (are) optically active?

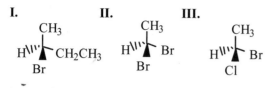

I.   II.   III.

A. I only
B. III only
C. II and III only
D. I and III only

135. In the assignment of absolute configuration, what priority does the Br atom have in the structure shown below?

A. 1
B. 2
C. 3
D. 4

136. In the assignment of absolute configuration, what priority does the ethyl group have in the structure shown below?

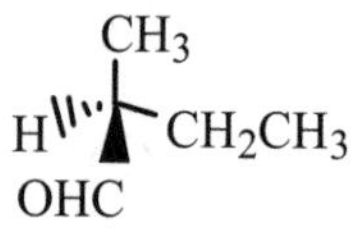

A. 1
B. 2
C. 3
D. 4

137. In the assignment of absolute configuration, what priority does the CHO group have in the structure shown below?

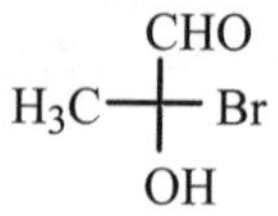

A. 1
B. 2
C. 3
D. 4

138. How are the pair of molecules shown below related?

A. diastereomers
B. enantiomers
C. epimers
D. not stereoisomers

139. How are the pair of molecules shown below related?

A. diastereomers
B. enantiomers
C. epimers
D. not stereoisomers

Refer to the structures of D-aldoses below that are synthesized from D-glyceraldhyde to answer questions 140–147.

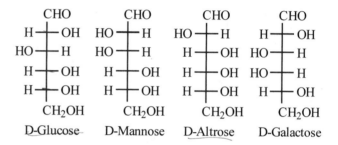

D-Glucose   D-Mannose   D-Altrose   D-Galactose

140. What is the most specific relationship between D-glucose and D-mannose?

A. diastereomers
B. enantiomers
C. epimers
D. not stereoisomers

141. What is the most specific relationship between D-glucose and D-altrose?

A. diastereomers
B. enantiomers
C. epimers
D. not stereoisomers

142. What is the most specific relationship between D-mannose and D-galactose?

A. diastereomers
B. enantiomers
C. epimers
D. not stereoisomers

143. What is the most specific relationship between D-mannose and D-altrose?

A. diastereomers
B. enantiomers
C. epimers
D. not stereoisomers

144. How many chiral centers are present in D-glucose?

A. 0
B. 1
C. 4
D. 6

145. What is the maximum number of optically active stereoisomers for the glucose molecule?

A. 0
B. 4
C. 8
D. 16

146. What is the relationship between D-glucose and the molecule shown below?

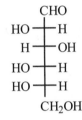

A. diastereomers
B. enantiomers
C. epimers
D. not stereoisomers

147. Of the four structures shown, which are diastereomers of D-glucose?

A. D-mannose
B. D-galactose and D-mannose
C. D-mannose, D-altrose, and D-galactose
D. D-altrose

Retinal is involved in the vision process. The significant transformation of retinal is shown below. Refer to this transformation to answer questions 148-150.

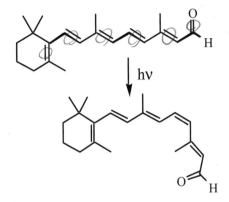

148. How many chiral centers are present in retinal?

A. 0
B. 1
C. 2
D. 3

149. What is the significant transformation that takes place in the retinal reaction?

A. conversion from one geometric isomer to another geometric isomer
B. conversion to the enantiomer
C. formation of an additional double bond
D. reduction of a double bond

150. How many double bonds are involved in the conjugated system of retinal?

A. 0
B. 4
C. 5
D. 6

Refer to the structure of cholesterol below to answer the questions 151-153.

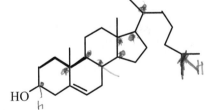

151. How many chiral centers are present in cholesterol?

A. 0
B. 2
C. 6
D. 8

152. What is the maximum number of optically active stereoisomers for the cholesterol molecule?

A. 4
B. 8
C. 16
D. 256

153. What geometric isomer is the cholesterol molecule?

A. trans
B. E
C. Z
D. The geometric isomer cannot be determined.

154. What is the absolute configuration of the malic acid molecule shown below?

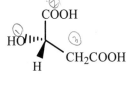

A. R
B. S
C. D
D. L

155. What is the absolute configuration of the lactic acid molecule shown below?

A. R
B. S
C. D
D. L

156. What direction does the molecule of lactic acid shown in questions 155 rotate light?

A. clockwise
B. counterclockwise
C. It cannot be determined without measurement
D. It does not rotate light.

157. What is the absolute configuration of the C2 and C5 carbons in D-glucose, respectively?

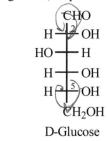

D-Glucose

A. R, R
B. R, S
C. S, R
D. S, S

The structure of naturally occurring epinephrine is shown below. Refer to this structure to answer questions 158-161.

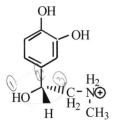

(l)-epinephrine

158. What is the absolute configuration of the chiral carbon in (l)-epinephrine?

A. R
B. S
C. D
D. L

159. What direction does (l)-epinephrine rotate plane polarized light?

   A. clockwise
   B. counterclockwise
   C. It cannot be determined without measurement
   D. It does not rotate light.

160. What is the relationship between the structure shown below and (l)-epinephrine?

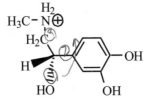

   A. diastereomers
   B. enantiomers
   C. structural isomers
   D. the same molecule

161. What direction does the structure shown in question 160 rotate plane polarized light?

   A. clockwise
   B. counterclockwise
   C. It cannot be determined without measurement
   D. It does not rotate light.

Carvone has two stereoisomers that are shown below. (+)-carvone smells like caraway seed and (-)-carvone has a spearmint odor. Refer to the carvone structures to answer questions 162-167.

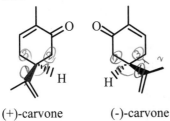

(+)-carvone   (-)-carvone

162. What geometric isomers are present in the (+)-carvone molecule?

   A. E, Z
   B. E, E
   C. Z for the first, and the second double bond does not have a geometric isomer.
   D. E for the first, and the second double bond does not have a geometric isomer.

163. What is the relationship between (+)-carvone and (-)-carvone?

   A. diasteromers
   B. enantiomers
   C. epimers
   D. not steroisomers

164. What are the absolute configurations of (+)-carvone and (-)-carvone respectively?

   A. R, R
   B. R, S
   C. S, R
   D. S, S

165. What direction does (+)-carvone rotate plane polarized light?

   A. clockwise
   B. counterclockwise
   C. cannot be determined without measurement
   D. the same direction as (-)-carvone

166. What physical properties will be different for (+)-carvone and (-)-carvone?

   I. Density
   II. Boiling point
   III. Rotation of plane polarize light

   A. I only
   B. III only
   C. I, II, and III
   D. I and III only

167. What is another name for (+)-carvone?

   A. (d)-carvone
   B. (l)-carvone
   C. D-carvone
   D. L-carvone

Refer to the reduction of 1,2-dibromopentene shown below to answer questions 168 - 173.

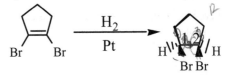

168. What is the geometric isomer present in 1,2-dibromopentene?

   A. E isomer
   B. Z isomer
   C. D isomer
   D. L isomer

169. What is the absolute configuration at C1 and C2 carbons respectively?

A. R, R
B. R, S
C. S, R
D. S, S

170. In what direction will the product of the 1,2-dibromopentene reaction rotate light?

A. clockwise
B. counterclockwise
C. cannot be determined without measurement
D. The molecule will not rotate light.

171. The 1,2-dibromopentane is called a(n):

A. chiral compound
B. anomeric compound
C. meso compound
D. enantiomer

172. What is the relationship between the structure shown below and the product from the reduction?

A. diastereomers
B. enantiomers
C. structural isomers
D. not stereoisomers

173. What is the name for the structure in question 172?

A. *cis*-1,2-dibromopentane
B. *trans*-1,2-dibromopentane
C. meso-1,2-dibromopentane
D. meso-dibromopentane

174. What are possible explanations for molecules with chiral centers not rotating plane polarized light?

I. There is a racemic mixture of the molecules.
II. The molecule is a meso compound.
III. Chiral molecules do not rotate light.

A. I only
B. III only
C. I and II only
D. I, II, and III

175. The enantiomers of a chiral alcohol are separated on a chiral column. This process is called:

A. racemization
B. resolution
C. rotation
D. meso

Refer to the structures of allene shown below to answer questions 176-178.

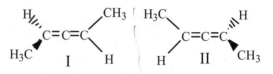

176. How many chiral centers are present in allene?

A. 0
B. 1
C. 2
D. 3

177. What is the relationship between structure I and structure II of allene?

A. diastereomers
B. enantiomers
C. epimers
D. not stereoisomers

178. What is the relationship between structure I of allene and the structure shown below?

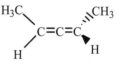

A. diastereomers
B. enantiomers
C. epimers
D. not stereoisomers

179. If S-leucine is found to rotate D line sodium light –11°, then R-leucine would rotate light D line sodium light:

A. 0°
B. +11°
C. –11°
D. –22°

180. A mixture of 2:1 (-)-leucine and (+)-leucine would rotate light

A. clockwise
B. counterclockwise
C. It cannot be determined without measurement.
D. It does not rotate light.

Tartaric acid was found to have distinctly different crystals. Two structures of tartaric acid are shown below. Refer to them to answer questions 181 - 189.

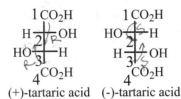

(+)-tartaric acid   (-)-tartaric acid

181. What is the absolute configuration, respectively of C2 and C3 of (+)-tartaric?

  A. R, R
  B. R, S
  C. S, R
  D. S, S

182. What is the absolute configuration, respectively of C2 and C3 of (-)-tartaric?

  A. R, R
  B. R, S
  C. S, R
  D. S, S

183. What is the relationship between (-)-tartaric and (+)-tartaric?

  A. diastereomers
  B. enantiomers
  C. epimers
  D. not stereoisomers

184. If (+)-tartaric acid is found to rotate D line sodium light +12°, then (-)-tartaric would rotate D line sodium light:

  A. 0°
  B. +12°
  C. −12°
  D. +24°

185. What is the relationship between (-)-tartaric and the structure shown below?

  A. diastereomers
  B. enantiomers
  C. structural isomer
  D. not stereoisomers

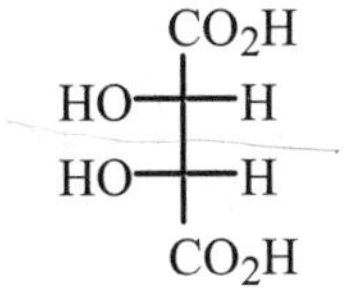

186. Which of the following is true about the structure shown in question 185?

  A. It has an enantiomer.
  B. It is a meso structure.
  C. It rotates light +12°.
  D. It rotates light -12°.

187. What is another name for (-)-tartaric acid?

  A. (d)-tartaric acid
  B. (l)-tartaric acid
  C. meso-tartaric acid
  D. (2R, 3R)-tartaric acid

188. How many optically active stereoisomers does tartaric acid have?

  A. 1
  B. 2
  C. 3
  D. 4

189. What is the best explanation for the fact that tartaric acid has two crystals that differ in appearance?

  A. (+)-tartaric acid and (-)-tartaric acid have different densities.
  B. (+)-tartaric acid and (-)-tartaric acid have different molecular weights.
  C. (+)-tartaric acid and (-)-tartaric acid have different melting points.
  D. (+)-tartaric acid and (-)-tartaric acid rotate plane polarized light in different directions.

190. In the substitution of a Cl atom for a bromine atom, the Cl atom adds to the opposite face from Br on the molecule. If the reaction begins with (S)-bromobutane, which is the correct structure of the product?

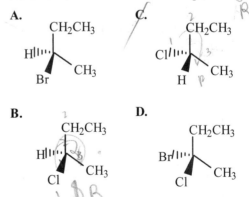

191. The reduction of 2-pentanone produces a racemic mixture of (R)-2-pentanol and (S)-2-pentanol. What direction will the mixture rotate light?

    A. clockwise
    B. counterclockwise
    C. It cannot be determined without measurement.
    D. The mixture will not rotate light.

192. A pure solution of 2-iodobutane rotates sodium D line light 15.90°. When reacted with methanol a racemic mixture of the ether product results. The product solution will rotate sodium D line light:

    A. 0°
    B. +15.90°
    C. −15.90°
    D. 31.80°

Only (l)-dopa can react with the enzymes in the brain to form dopamine. The reaction is shown below. Refer to the reaction shown below to answers questions 193-200.

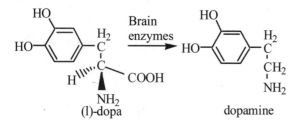

193. The conversion that takes place in the brain when (l)-dopa is converted to dopamine is a(n):

    A. decarboxylation
    B. reduction of an amide to an amine
    C. oxidation of an amide to an amine
    D. carboxylation

194. The absolute configuration of the chiral carbon in (l)-dopa is:

    A. R
    B. S
    C. D
    D. L

195. Another name for (l)-dopa is

    A. (+)-dopa
    B. (-)-dopa
    C. (d)-dopa
    D. (R)-dopa

196. What direction will (l)-dopa rotate light?

    A. clockwise
    B. counterclockwise
    C. cannot be determined without measurement
    D. The molecule will not rotate light.

197. How many chiral centers does dopamine have?

    A. 0
    B. 1
    C. 2
    D. 3

198. What direction will dopamine rotate light?

    A. clockwise
    B. counterclockwise
    C. It cannot be determined without measurement.
    D. The molecule will not rotate light.

199. The enantiomer of (l)-dopa is toxic. Which of the following structure is the enantiomer of (l)-dopa?

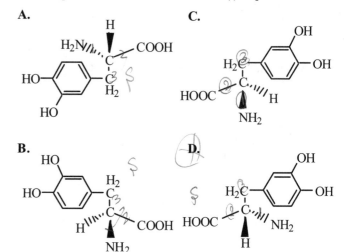

200. If a scientist needed to separate (l)-dopa from (d)-dopa, what processes would work to accomplish this resolution?

    I. Distillation
    II. Recrystallization
    III. Formation of diastereomers with an enantiomerically pure compound.

    A. I only
    B. III only
    C. I, II, and III
    D. I and III only

Isopulegols result as a side product from the oxidation of citronellol. Refer to the structure below to answer questions 201-204.

Isopulegols

201. How many chiral centers are present in isopulegols?

- **A.** 0
- **B.** 1
- **C.** 2
- **D.** 3

202. What is the maximum number of optically active stereoisomers for the isopulegols?

- **A.** 2
- **B.** 3
- **C.** 4
- **D.** 8

203. Pulegone can be synthesized from isopulegone. The structure of pulegone is shown below. How many chiral centers does pulegone have?

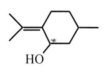

- **A.** 0
- **B.** 1
- **C.** 2
- **D.** 3

204. Which of the following structures is not a stereoisomer of the isomer of isopulegol shown below?

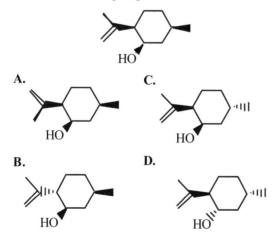

205. A neat solution of 1-phenylethanol is optically active, while 2-phenylethanol is not. The best explanation for this observation is

- **A.** 2-phenylethanol has a higher molecular weight than 1-phenylethanol.
- **B.** 2-phenylethanol has a higher boiling point than 1-phenylethanol.
- **C.** 1-phenylethanol is chiral, while 2-phenylethanol iss not.
- **D.** 2-phenylethanol is chiral, while 1-phenylethanol is not.

206. The structures of (+)-1-phenylethanol and (-)-1-phenylethnaol are shown below. What is the absolute configuration of the (+) and (-) isomer, respectively?

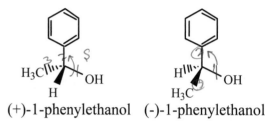

(+)-1-phenylethanol   (-)-1-phenylethanol

- **A.** R, R
- **B.** S, S
- **C.** S, R
- **D.** R, S

207. Given that a neat sample of (+)-1-phenylethanol rotates the sodium D line +42°, what can be determined from the observation that a pure neat sample of 1-phenylethanol rotates the sodium D line +12°?

- **A.** The sample contains a mixture of R and S isomers, with more R isomer present.
- **B.** The sample contains a mixture of R and S isomers, with more S isomer present.
- **C.** The sample contains pure R isomer.
- **D.** The sample contains pure S isomer.

208. Yeast selectively reduces the ketone carbonyl in ethyl acetoacetate to S(+)-ethyl-3-hydroxybutanoate. Which of the following is the product of the reaction shown below?

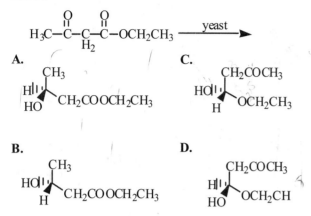

A.

CH₃
H| |
HO   CH₂COOCH₂CH₃

C.

CH₂COCH₃
HO| |
H   OCH₂CH₃

B.

CH₃
HO| |
H   CH₂COOCH₂CH₃

D.

CH₂COCH₃
H| |
HO   OCH₂CH

209. In what direction would the ethylacetoacetate and the S(+)-ethyl-3-hydroxybutanoate (discussed in question 208) rotate plane polarized light respectively?

A. clockwise then counterclockwise
B. counterclockwise then clockwise
C. no rotation then clockwise
D. no rotation then counter clockwise

Alkanes

210. Which of the following structures is an alkane?

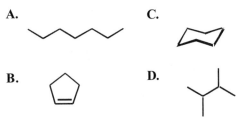

A.

C.

B.

D.

211. Which of the following is <u>not</u> an alkane?

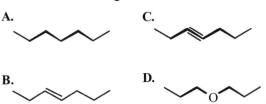

A.

C.

B.

D.

212. Which of the following could <u>not</u> be an alkane?

A. C_4H_{10}
B. C_6H_{12}
C. $C_{10}H_{22}$
D. $C_{10}H_{22}O$

213. Which of the following is a haloalkane?

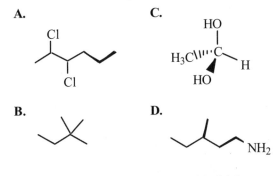

A.

C.

B.

D.

214. Which of the following is a geminal halide?

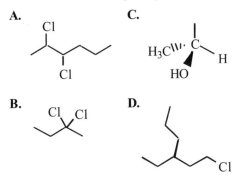

A.

C.

B.

D.

Refer to the structure of 4-methylheptane shown below to answer questions 215-217.

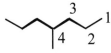

215. The carbon labeled C1 in 4-methylheptane is a

A. methyl carbon
B. primary carbon
C. secondary carbon
D. tertiary carbon

216. The carbon labeled C4 in 4-methylheptane is a

A. methyl carbon
B. primary carbon
C. secondary carbon
D. tertiary carbon

217. The carbon labeled C3 in 4-methylheptane is a

A. methyl carbon
B. primary carbon
C. secondary carbon
D. tertiary carbon

Refer to the structure of 3β-cholestanol to answer questions 218-219.

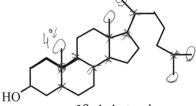

3β-cholestanol

218. How many tertiary carbons are in 3β-cholestanol?

A. 3
B. 5
C. 7
D. 9

219. How many primary carbons are in 3β-cholestanol?

A. 3
B. 5
C. 7
D. 9

220. Which of the following is a radical?

A.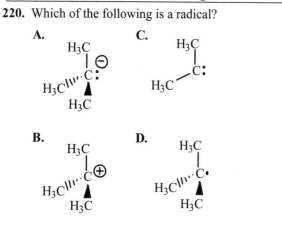
C.

B.
D.

221. Which of the following radicals is the most stable?

A.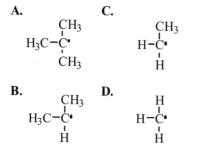
C.

B.
D.

222. Which of the following radicals is the least stable?

A.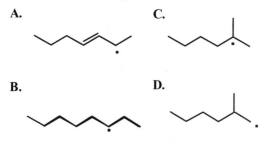
C.

B.
D.

223. In the structure shown below which carbon would be most reactive in a radical reaction?

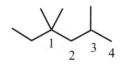

A. carbon 1
B. carbon 2
C. carbon 3
D. carbon 4

224. The IUPAC name for the structure shown below is:

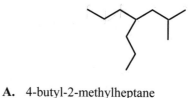

A. 4-butyl-2-methylheptane
B. 2-methyl-4-butylheptane
C. 6-methyl-4-propylheptane
D. 2-methyl-4-propylheptane

225. The IUPAC name for the structure shown below is:

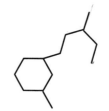

A. 1-(3-methylpentyl)-3-methylcyclohexane
B. 1-(2-ethylbutyl)-3-methylcyclohexane
C. 2-ethyl-7-methyldodecane
D. 7-ethyl-2-methyldodecane

226. Which of the following alkanes has the highest boiling point?

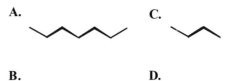

A.
C.

B.
D.

227. Which of the following alkanes has the lowest boiling point?

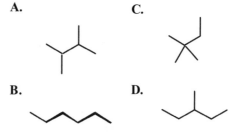

A.
C.

B.
D.

228. Which of the following alkanes has the lowest density?

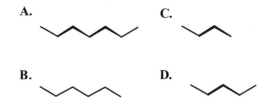

A.
C.

B.
D.

229. Which of the following alkanes has the lowest melting point?

A. nonane
B. octane
C. butane
D. dodecane

230. Combustion is what type of reaction?

A. exothermic
B. endothermic
C. kinetically favored
D. nonspontaneous

231. Which of the following are steps in radical halogenation?

I. initiation
II. propagation
III. termination

A. II only
B. I only
C. I and II only
D. I, II, and III

232. Which of the following compounds is least reactive?

A. diethyl ether
B. heptanal
C. butane
D. acetic anhydride

233. Heptane has a boiling point of 98°C. Which alkane would have a higher boiling point?

A. pentane
B. octane
C. butane
D. methane

234. Nonane has a boiling point of 151°C. Which alkane would have a higher boiling point?

A. dodecane
B. octane
C. 3-methyloctane
D. 2,2-dimethylheptane

235. When an alkane is reacted with oxygen at high temperatures, what are the products?

A. CO_2
B. H_2O
C. CO_2 and H_2O
D. CO_2 and H_2

236. In order to achieve combustion of an alkane, what is required?

A. O_2
B. O_2 and high temperatures
C. H_2 and high pressure
D. CO_2 and H_2O

237. A series of alkanes is separated by distillation. Which alkane would be the first to boil?

A. pentane
B. octane
C. hexane
D. heptane

238. Given that the boiling point of pentane is 36°C, what state is octane in at room temperature?

A. gas
B. liquid
C. solid
D. The state cannot be determined.

239. Given that the density of pentane is 0.56 and the density of octane is 0.70, what is the density of hexane?

A. 0.44
B. 0.56
C. 0.66
D. 0.70

240. Given that the boiling point of butane is 0°C, what state is propane in at room temperature?

A. gas
B. liquid
C. solid
D. The state cannot be determined.

241. How many moles of H_2 are produced when methane is completely combusted?

A. 0
B. 2
C. 4
D. 5

242. How many moles of H_2O are produced when butane is completely combusted?

A. 0
B. 2
C. 4
D. 5

243. How many moles of H_2O are produced when cyclooctane is completely combusted?

A. 0
B. 4
C. 8
D. 16

244. How many moles of CO_2 are produced when heptane is completely combusted?

A. 0
B. 1
C. 7
D. 14

245. How many moles of CO_2 are produced when cyclopropane is completely combusted?

A. 0
B. 1
C. 2
D. 3

246. How many moles of O_2 are consumed when nonane is completely combusted?

A. 4.5
B. 7
C. 14
D. 28

247. How many moles of O_2 are consumed when cyclohexane is completely combusted?

A. 6
B. 9
C. 12
D. 15

248. Which of the following alkanes would have the largest heat of combustion?

A. methane
B. butane
C. octane
D. cyclohexane

249. Given that decane has a melting point of -30°C and hexane has a melting point of -95°C, what is the melting point of octane?

A. -25°C
B. -57°C
C. -95°C
D. -100°C

250. Given that butane has a boiling point of 0°C and pentane has a boiling point of 36°C, what is the boiling point of isopentane?

A. -10°C
B. 30°C
C. 36°C
D. 45°C

251. Given that cyclopentane has a boiling point of 49°C and cyclooctane has a boiling point of 148°C, what is the boiling point of cyclohexane?

A. 36°C
B. 49°C
C. 81°C
D. 150°C

252. Which of the following places the densities for the alkanes from lowest to highest?

A. ethane < heptane < hexane< octane
B. ethane < hexane < heptane< octane
C. ethane < heptane < hexane< pentane
D. methane < heptane < hexane< ethane

253. Which of the following places the boiling points for the alkanes from lowest to highest?

A. heptane < hexane < t-butane< octane
B. isobutane < butane < isopentane< pentane
C. butane < t-butane < heptane < hexane
D. butane < isobutane < pentane< isopentane

254. What is the conformation of cyclohexane shown below called?

A. boat
B. twisted boat
C. half chair
D. chair

255. What is the conformation of cyclohexane shown below called?

A. boat
B. twisted boat
C. half chair
D. chair

256. Which of the following is the most stable conformation of cyclohexane?

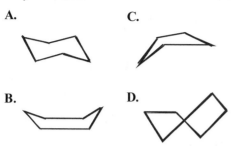

A.

C.

B.

D.

257. Which of the following structures has the lowest ring strain?

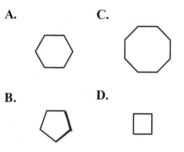

A.

C.

B.

D.

258. Alkanes are fairly unreactive compounds. What reactions do alkanes undergo?

   I.   Radical reactions
  II.   Substitution
 III.   Combustion

   A.   I only
   B.   II only
   C.   I and II only
   D.   I and III only

259. If a scientist had a long chain alkane to dissolve, what solvent would be the best choice?

   A. H_2O
   B.   ethanol
   C.   heptane
   D.   ethyl acetate

260. The best explanation for the fact that decane has a boiling point of 174°C and pentane has a boiling point of 36°C is

   A.   decane has stronger hydrogen bonding than pentane.
   B.   pentane has stronger hydrogen bonding than decane.
   C.   decane has stronger London dispersion forces than pentane.
   D.   pentane has stronger London dispersion forces than decane.

261. What is the best explanation for the trend in boiling points observed in the table below?

	Boiling point
2-methylpentane	60 °C
2,3-dimethylbutane	58 °C
2,2-dimethylbutane	50 °C

   A.   As branching increases London dispersion forces decrease.
   B.   As branching increases London dispersion forces increase.
   C.   As molecular weight increases London dispersion forces increase.
   D.   As molecular weight increases London dispersion forces decrease.

262. The boiling point of $CH_3CH_2CH_2F$ is 3°C, while the boiling point of $CH_3CH_2CH_3$ is -42°C. Which of the following accounts for the dramatic difference in boiling point despite their similar size?

   A.   1-Fluoropropane has hydrogen bonding.
   B.   1-Fluoropropane has stronger London dispersion forces.
   C.   1-Fluoropropane has dipole-dipole interactions.
   D.   1-Fluoropropane has weaker London dispersion forces.

263. Paraffin "wax" is composed of solid alkanes. Which of the following alkanes would be the best choice use in the manufacture of paraffin.

   A. CH_4
   B. C_5H_{12}
   C. C_4H_{10}
   D. $C_{25}H_{52}$

264. In cold weather, diesel fuel, which is composed of 14 carbon alkanes, freezes. Which of the following would work best as an additive to diesel fuel to prevent freezing?

   A. H_2O
   B.   ethanol
   C.   gasoline (composed of pentanes through octanes)
   D.   mineral oil (composed of 16 to 18 carbon alkanes)

265. Natural gas is a byproduct of the production of petroleum. Before the value of this byproduct was known, it was disposed of by flaring (burning the gas). What products were produced during flaring?

   A. O_2
   B. H_2O
   C. CO_2 and H_2O
   D. CO_2 and H_2

266. Which of the following steps is an initiation step?

    A. $Cl_2 + h\nu \longrightarrow 2Cl\cdot$

    B. $Cl\cdot + CH_3CH_3 \longrightarrow HCl + CH_3CH_2\cdot$

    C. $CH_3CH_3 + CH_3CH_2\cdot \longrightarrow CH_3CH_3 + CH_3CH_2\cdot$

    D. $CH_3CH_2\cdot + Cl\cdot \longrightarrow CH_3CH_2Cl$

267. Which of the following steps is a termination step?

    A. $Cl_2 + h\nu \longrightarrow 2Cl\cdot$

    B. $Cl\cdot + CH_3CH_3 \longrightarrow HCl + CH_3CH_2\cdot$

    C. $CH_3CH_3 + CH_3CH_2\cdot \longrightarrow CH_3CH_3 + CH_3CH_2\cdot$

    D. $CH_3CH_2\cdot + Cl\cdot \longrightarrow CH_3CH_2Cl$

268. Which of the following steps requires a homolytic cleavage?

    I. initiation
    II. propagation
    III. termination

    A. II only
    B. I only
    C. I and II only
    D. I, II, and III

269. Which of the following steps is not a propagation step in a radical reaction?

    A. $Cl_2 + CH_3CH_2\cdot \longrightarrow CH_3CH_2Cl + Cl\cdot$

    B. $Cl\cdot + CH_3CH_3 \longrightarrow HCl + CH_3CH_2\cdot$

    C. $CH_3CH_3 + CH_3CH_2\cdot \longrightarrow CH_3CH_3 + CH_3CH_2\cdot$

    D. $CH_3CH_2\cdot + Cl\cdot \longrightarrow CH_3CH_2Cl$

270. Which of the following steps always generates a radical in the product?
    I. initiation
    II. propagation
    III. termination

    A. II only
    B. I only
    C. I and II only
    D. I, II, and III

271. Why is ultraviolet light required for radical reactions?

    A. Ultraviolet light provides energy to break all the bonds in the reaction.
    B. Ultraviolet light is required to form the product bond in the termination step.
    C. Ultraviolet light is required to break the carbon-hydrogen bond in the alkane.
    D. Ultraviolet light provides energy to break a bond and initiate the reaction.

272. Which of the bonds labeled in the structure below would require the least amount of energy to break?

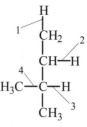

    A. the bond labeled 1
    B. the bond labeled 2
    C. the bond labeled 3
    D. the bond labeled 4

273. What is the order from most to least reactive for the halogen radicals?

    A. $F > Cl > Br > I$
    B. $Cl > Br > F > I$
    C. $Br > Cl > F > I$
    D. $I > Br > Cl > F$

274. When Cl_2 is reacted with iso-butane, the chlorine radical reacts 5.5 times faster with the tertiary hydrogen. What is the best explanation for the distribution of products shown below?

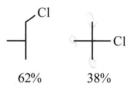

    62%   38%

    A. The distribution of products is unrelated to relative reactivities.
    B. There are nine primary hydrogens compared to one tertiary hydrogen.
    C. Primary halogens are more stable than tertiary halogens.
    D. The selectivity of chlorine radicals is greater than bromine radicals.

275. Based the bond dissociation energies in the table below, which reagent would be best to use for the initiation step of a radical reaction?

Bond	Bond Dissociation Energy
HO - H	119 kcal/mol
HO - OH	51 kcal/mol
CH_3 - F	109 kcal/mol
$(CH_3)_3C$ - H	91 kcal/mol
$CH_3 - CH_3$	88 kcal/mol

    A. H_2O_2
    B. H_2O
    C. CH_3F
    D. C_2H_6

276. The bond energies shown below indicate that the allyl radical is more stable than the butyl radical. What is the best explanation for this observation?

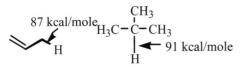

A. The butyl radical is an unstable radical.
B. The allyl radical is an unstable radical.
C. Steric hindrance on the t-butyl group destabilizes the bond.
D. The allyl radical is resonance stabilized.

277. The hydrogen shown on the cyclohexane ring below is a(n):

A. axial hydrogen
B. equatorial hydrogen
C. cis hydrogen
D. trans hydrogen

278. The hydrogen shown on the cyclohexane ring below is a(n):

A. axial hydrogen
B. equatorial hydrogen
C. cis hydrogen
D. trans hydrogen

279. The geometric relationship between the two methyl groups shown on the cyclohexane ring below is:

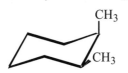

A. axial
B. equatorial
C. cis
D. trans

280. What is the most stable confirmation of *cis*-1-t-butyl-3-methylcyclohexane?

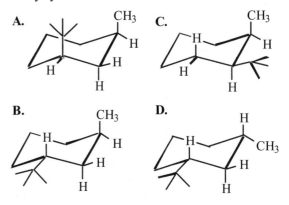

281. What is the most stable conformation of the cyclohexane ring shown below?

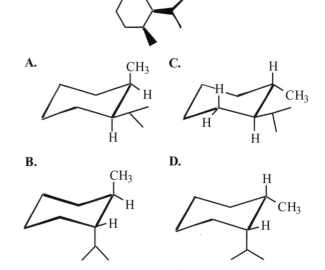

282. What is the most stable confirmation of the cyclohexane ring shown below?

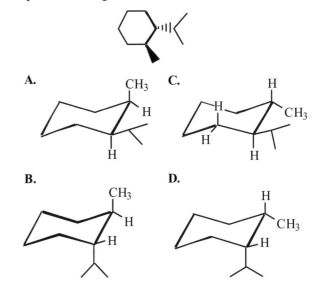

283. What is the best explanation for the trend observed in the table?

Radical Reacted with CH_4	Relative Rate ($e^{-Ea/RT}$ x 10^6) at 27°C
F	140,000
Cl	1300
Br	9×10^{-8}
I	2×10^{-19}

A. As the atomic mass of the halogen increases, the rate of reaction increases.
B. As the atomic radius of the halogen increases, the rate of reaction increases.
C. As the reactivity of the halogen radical decreases, the rate of reaction increases.
D. As the reactivity of the halogen radical increases, the rate of reaction increases.

284. How many monochlorinated products are possible when Cl_2 is treated with UV light and reacted with 3-methylpentane?

A. 0
B. 1
C. 4
D. 8

285. How many monobrominated products result when Br_2 is treated with UV light and reacted with 3-methylpentane?

A. 0
B. 1
C. 4
D. 8

286. NBS produces a low concentration of bromine radicals. What is the monobrominated product when NBS is reacted with 2,3-dimethylbutane?

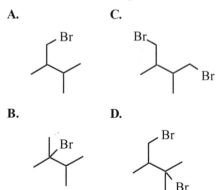

A.
C.
B.
D.

287. Which of the following are possible monochlorinated products, when Cl_2 is reacted with 3,5-dimethylheptane?

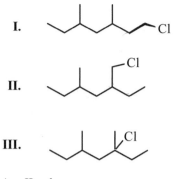

I.

II.

III.

A. II only
B. I only
C. I and II only
D. I, II, and III

288. Radical halogenation has limited use in industry to produce haloalkanes. The best explanation for this fact is that

A. radicals are too expensive to produce.
B. radical reactions are difficult to control.
C. radicals are difficult to produce on a large scale.
D. the only source of radicals requires large amounts of heat.

289. What will be the major monofluorinated product when 2-methylbutane undergoes a free-radical reaction with F_2?

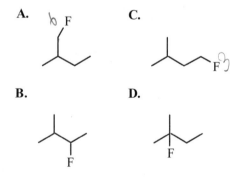

A.
C.
B.
D.

290. The best explanation for the stability of the chair confirmation for cyclohexane is:

I. the bond angles are 109.5°
II. the hydrogens are all in staggered conformations
III. the bonds have nonlinear overlap

A. I only
B. II only
C. I and II only
D. I, II, and III

291. What is the best explanation for the trend observed in the table below?

Cycloalkane	Heat of combustion per CH_2 group
cyclopropane	166.6
cyclobutane	164.0
cyclopentane	158.7
cyclohexane	157.4
cycloheptane	158.3
cyclooctane	158.6

  A. The heat of combustion per CH_2 decreases as the number of CH_2 groups increases.
  B. The heat of combustion per CH_2 increases as the number of CH_2 groups increases.
  C. The heat of combustion per CH_2 increases as ring strain increases.
  D. The heat of combustion per CH_2 increases as ring strain decreases.

292. Combustion reactions are difficult to control. One method of control is to regulate the fuel/air ratio. What does regulation of this ratio control?

  A. the types of reactants
  B. the types of product
  C. the ratio of reactants to products
  D. the ratio of reactants

Alkenes

293. Which of the following structures is an alkene?

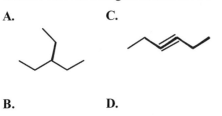

  A.   **C.**

  B.   **D.**

294. Which of the following structures is an alkyne?

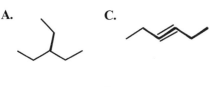
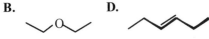

  A.   **C.**

  B.   **D.**

295. Which of the following alkenes is the most stable?

  A.   **C.**

  B.   **D.**

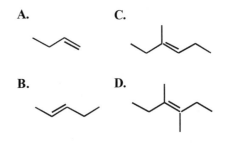

296. Which of the following alkenes is the least stable?

  A.   **C.**

  B.   **D.**

297. What is the correct IUPAC name for the alkene shown below?

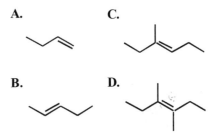

  A. 2-metylbutyl-2-butene
  B. 2-metylbutyl-3-butene
  C. 3-ethyl-5-methyl-2-heptene
  D. 3-ethyl-5-methyl-3-heptene

298. What is the correct IUPAC name for the alkyne shown below?

  A. 4,4-dimethyl-2-hexyne
  B. 2,2-dimethyl-4-hexyne
  C. 2,2-dimethyl-4-hexene
  D. 4-ethyl-4-methyl-2-pentyne

299. What is the correct name for the structure shown below?

  A. ortho-nitrotoluene
  B. meta-nitrotoluene
  C. para-nitrotoluene
  D. para-toluene

300. What is another name for 4-methoxybenzoate?

    A. p-anisate
    B. m-anisate
    C. o-anisate
    D. m-toluate

301. For the structure below, give the geometric assignment for the bond between C2 and C3 and the bond between C4 and C5, respectively.

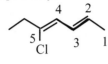

    A. E, E
    B. Z, Z
    C. E, Z
    D. Z, E

302. For the structure below, give the geometric assignment for the bond between C1 and C2 and the bond between C5 and C6, respectively.

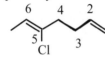

    A. neither, E
    B. neither, Z
    C. E, Z
    D. Z, E

303. Alkenes are more acidic than alkanes. What is the best explanation for this trend?

    A. The sigma bond present in alkenes helps to stabilize the negative charge generated when a proton is removed.
    B. The sigma bond present in alkenes helps to destabilize the negative charge generated when a proton is removed.
    C. The pi bond present in alkenes helps to stabilize the negative charge generated when a proton is removed.
    D. The pi bond present in alkenes helps to destabilize the negative charge generated when a proton is removed.

304. Given that 1-butene has a boiling point of -6°C and 1-hexene has a boiling point of 64°C, what is the boiling point of 1-pentene?

    A. -10°C
    B. -6°C
    C. 30°C
    D. 74°C

305. Given that benzene molecules with symmetry pack better in a crystalline structure raising the melting point, which of the following would have the greatest melting point?

    A. o-dichlorobenzene
    B. p-dichlorobenzene
    C. m-dichlorobenzene
    D. All three would have similar melting points.

306. What is the best explanation for the fact that 1-butene has a boiling point of –6.3°C and 1-butyne has a boiling point of 8.1°C?

    A. 1-Butene has a higher molecular weight than 1-butyne.
    B. 1-Butyne has a higher molecular weight than 1-butene.
    C. 1-Butyne is more polar than 1-butene.
    D. 1-Butene is more polar than 1-butene.

307. Given that 1-butyne has a boiling point of 8.1°C, what is the phase of propyne at room temperature and 1 atm pressure?

    A. gas
    B. liquid
    C. solid
    D. supercritial fluid

308. Which of the following carbocations is the most stable?

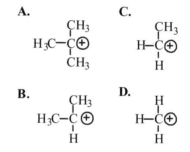

309. Which of the following is the least stable cation?

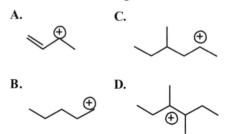

310. Which of the following is an electron withdrawing group when attached to benzene?

- **A.** CH_3
- **B.** Cl
- **C.** OH
- **D.** OCH_3

311. Which of the following is an electron donating group when attached to benzene?

- **A.** NO_2
- **B.** CO_2H
- **C.** Cl
- **D.** OCH_3

312. What would be the resulting product if one mole of 3-hexyne were hydrogenated with 1 mole of H_2?

- **A.** hexane
- **B.** *cis*-3-hexene
- **C.** *trans*-3-hexene
- **D.** 1-hexene

313. What would be the resulting product if one mole of 3-hexyne were hydrogenated with an excess of H_2?

- **A.** hexane
- **B.** *cis*-3-hexene
- **C.** *trans*-3-hexene
- **D.** 1-hexene

314. What conditions are required for hydrogenation of an alkene?

- **A.** metal catalyst
- **B.** metal catalyst and high temperatures
- **C.** metal catalyst and low temperatures
- **D.** low temperatures and high pressures

315. Which of the following can be used to catalyze hydrogenation?

- **I.** Ni
- **II.** Pt
- **III.** H^+

- **A.** I only
- **B.** II only
- **C.** I and II
- **D.** I and III

316. How many moles of hydrogen are required to convert a mole of pentyne to pentane?

- **A.** 0
- **B.** 1
- **C.** 2
- **D.** 3

A lower heat of hydrogenation indicates a more stable alkene. Refer to the table below to answer questions 317-322.

Alkene	Molar heat of hydrogenation
ethylene	137 kJ
propene	126 kJ
1-butene	127 kJ
cis-2-butene	120 kJ
trans-2-butene	116 kJ
cis-2-pentene	120 kJ
2-methyl-2-butene	113 kJ
2,3-dimethyl-2-butene	111 kJ

317. Which of the following alkenes is the most stable?

- **A.** ethylene
- **B.** cis-2-butene
- **C.** 2-methyl-2-butene
- **D.** 2,3-dimethyl-2-butene

318. Which of the following alkenes is the least stable?

- **A.** ethylene
- **B.** cis-2-butene
- **C.** trans-2-butene
- **D.** 2,3-dimethyl-2-butene

319. What would be the predicted molar heat of hydrogenation for trans-2-pentene?

- **A.** 111 kJ
- **B.** 116 kJ
- **C.** 120 kJ
- **D.** 124 kJ

320. The data in the table supports which of the following statements.

- **A.** Stability of an alkene increases with molecular weight.
- **B.** Stability of an alkene decreases with molecular weight.
- **C.** Stability of an alkene increases with substitution on the alkene carbons.
- **D.** Stability of an alkene decreases with substitution on the alkene carbons.

321. Why does *cis*-2-butene have a molar heat of hydrogenation of 120 kJ, while *trans*-2-butene has one of 116 kJ?

A. The cis isomer is more stable than the trans due to steric interactions.
B. The trans isomer is more stable than the cis due to steric interactions.
C. The double bond in the trans isomer has more potential energy.
D. Trans-2-butene has a higher molecular weight.

322. Why does *cis*-4,4-dimethyl-2-pentene have a heat of hydrogenation that is 4 kcals greater than *trans*-4,4-dimethyl-2-pentene, while the heat of hydrogenation of *cis*-2-pentene is only 1 kcal greater than that of *trans*-2-pentene?

A. Cis and trans isomers always differ by 1 kcal of energy.
B. Cis isomers always have greater heats of hydrogenation than trans isomers.
C. Because 4,4-dimethyl-2-pentene has bulkier groups, the steric effect is greater leading to a smaller difference in energy between the isomers.
D. Because 4,4-dimethyl-2-pentene has bulkier groups, the steric effect is greater leading to a greater difference in energy between the isomers.

323. To completely hydrogenate benzene to cyclohexane requires H_2, an Rh catalyst, and 1000 psi pressure at 100°C. Why are these conditions required?

A. The double bonds in benzene are more reactive than a typical alkene.
B. The double bonds in benzene are less reactive than a typical alkene.
C. The double bonds in benzene have the same reactivity as a typical alkene.
D. Hydrogenation produces an aromatic compound.

324. An alkene that has the formula $C_{10}H_{18}$ is hydrogenated with an excess of H_2, and the resulting alkane has the formula $C_{10}H_{22}$. How many double bonds were present in the original alkene?

A. 0
B. 1
C. 2
D. 3

325. An alkene that has the formula C_8H_{12} requires 3 moles of H_2 per mole of alkene to form a saturated alkane. What is the formula of the alkane?

A. C_8H_{15}
B. C_8H_{16}
C. C_8H_{18}
D. C_8H_{20}

326. Oil of celery is an alkene with the molecular formula $C_{15}H_{24}$. After hydrogenation with an excess of H_2 the molecular formula is $C_{15}H_{28}$. What can be concluded about the structure of oil of celery?

A. Oil of celery has two double bonds and two rings.
B. Oil of celery has four double bonds and no rings.
C. Oil of celery has no double bonds and four rings.
D. Nothing can be determined about the structure.

327. What does the letter E represent in the E1 notation?

A. elimination
B. excess of reagent
C. energy
D. the "ene" in alkene

328. What does the number 1 represent in the E1 notation?

A. number of mechanistic steps
B. order of the rate law
C. number of bonds formed
D. number of bonds broken

329. In an E1 reaction, what are the bond conversions in the major product?

A. 2 sigma bonds are converted to 1 pi bond.
B. 1 sigma bond is converted into 2 pi bonds.
C. 1 pi bond is converted into 2 sigma bonds.
D. 2 pi bonds are converted into 1 sigma bond.

330. In electrophilic addition to an alkene, what are the bond conversions in the major product?

A. 2 sigma bonds are converted to 1 pi bond.
B. 1 sigma bond is converted into 2 pi bonds.
C. 1 pi bond is converted into 2 sigma bonds.
D. 2 pi bonds are converted into 1 sigma bond.

331. Which of the following is an electrophile?

A. OH^-
B. NH_3
C. H^+
D. CH_4

332. Why does benzene undergo substitution reactions rather than addition reactions?

A. Benzene does not have a double bond.
B. If benzene underwent an addition reaction, the aromaticity of the ring would be disrupted.
C. When benzene undergoes a substitution reaction, the aromaticity is disrupted.
D. Benzene is an alkene.

333. If 2-methyl-2-butanol is heated with sulfuric acid, what are the resulting products?

   I. 2-methyl-1-butene
   II. 2-methyl-2-butene
   III. 3-methyl-1-butene

   A. I only
   B. III only
   C. I and II only
   D. I, II, and III

334. What is the purpose of the sulfuric acid in the reaction discussed in question 333?

   A. source of sulfate ion
   B. catalyst
   C. control pH
   D. source of hydroxide ions

335. What is the product when 2-methyl-2-pentene is reacted with HBr?

   A. 2-bromo-2-methylpentane
   B. 3-bromo-2-methylpentane
   C. 2-bromo-pentane
   D. 2-bromo-2-methylpentene

336. If 2-butanol is heated with sulfuric acid, what are the resulting products?

   I. 1-butene
   II. *cis*-2-butene
   III. *trans*-2-butene

   A. I only
   B. III only
   C. I and II only
   D. I, II, and III

337. Which of the following will oxidize an alkene?

   A. concentrated H^+ and heat
   B. H_2 with Ni
   C. O_3 and $(CH_3)_2S$
   D. O_2 and heat

338. Which of the following will reduce an alkyne?

   A. concentrated H^+ and heat
   B. H_2 with Ni ✓
   C. O_3 and $(CH_3)_2S$
   D. O_2 and heat

339. Which of the following will promote hydration of an alkene to an alcohol?

   A. concentrated H^+ and heat
   B. H_2 with Ni
   C. O_3 and $(CH_3)_2S$
   D. dilute acid and cold conditions

340. Which of the following alkenes would react the fastest with HBr?

   A. ethene
   B. 1-butene
   C. 2-butene
   D. 2-methyl-2-butene

341. Which of the following compounds would be most reactive when Cl_2 and $AlCl_3$ are added?

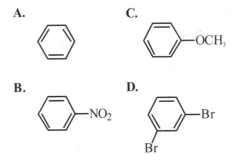

342. Which of the following compounds would be least reactive when Br_2 and $FeBr_3$ are added?

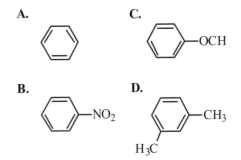

343. Which of the following is not a possible product of an ozonolysis?
   I. aldehyde
   II. ketone
   III. alcohol

   A. I only
   B. III only
   C. I and II only
   D. I, II, and III

344. What is the product of the following reaction?

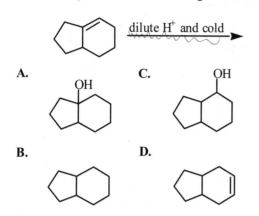

dilute H⁺ and cold

A.

C.

B.

D.

345. What is the product of the following reaction?

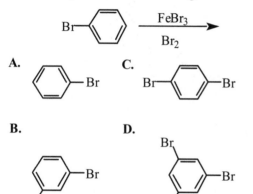

FeBr₃

Br₂

A.

C.

B.

D.

346. What is the product of the following reaction?

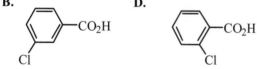

AlCl₃

Cl₂

A.

C.

B.

D.

347. What is the product of the following reaction?

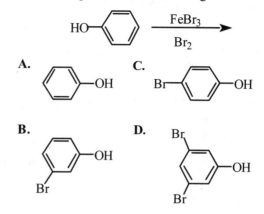

FeBr₃

Br₂

A.

C.

B.

D.

348. What is the product of the following reaction?

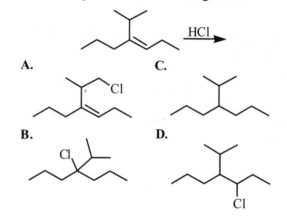

HCl

A.

C.

B.

D.

349. What is the purpose of FeBr₃ in the aromatic substitution of Br?

A. to provide an activated electrophile
B. to provide an activated nucleophile
C. to be a base
D. to terminate the reaction

Myrcene, a terpene found in bayberry is shown below. Use this structure to answer questions 350 - 353.

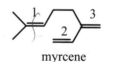

myrcene

350. Which of the double bonds in myrcene is the most stable?

A. The bond labeled 1.
B. The bond labeled 2.
C. The bond labeled 3.
D. All the bonds have the same stability.

351. How many moles of H_2 would be required to hydrogenate myrcene to a saturated alkane?

   A. 1
   B. 3
   C. 6
   D. 5

352. How many geometric isomers of myrcene exist?

   A. 0
   B. 3
   C. 6
   D. 9

353. Which of the following is <u>not</u> a product of the complete ozonolysis of mycrene?

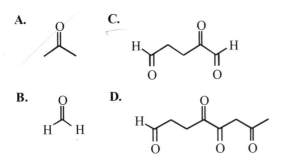

354. Which of the following alcohols would undergo dehydration by an E1 mechanism the fastest?

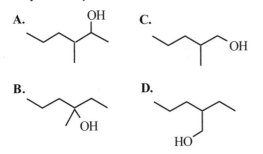

Refer to the structure of cholesterol to answer questions 355 - 358.

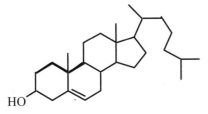

355. What reagents could be used to covert cholesterol to the following structure?

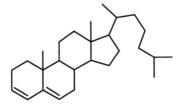

   A. concentrated H^+ and heat
   B. H_2 with Ni
   C. O_3 and $(CH_3)_2S$
   D. dilute acid and cold conditions

356. What reagents could be used to covert cholesterol to the following structure?

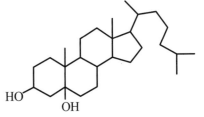

   A. concentrated H^+ and heat
   B. H_2 with Ni
   C. O_3 and $(CH_3)_2S$
   D. dilute acid and cold conditions

357. What reagents could be used to covert cholesterol to the following structure?

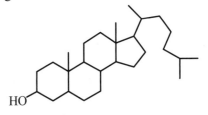

   A. concentrated H^+ and heat
   B. H_2 with Ni
   C. O_3 and $(CH_3)_2S$
   D. dilute acid and cold conditions

358. What would be the product when cholesterol is reacted with Br₂ in the presence of CH₃COOH?

A.

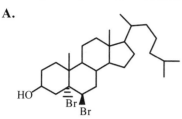

B.

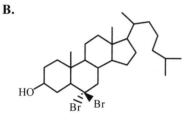

C.

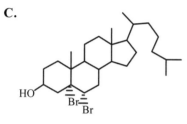

D.

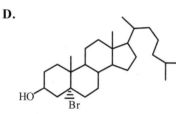

A terpene that is contained in the oil of citronella is α-farnesene. Refer to the structure of α-farnesene to answer questions 359-365.

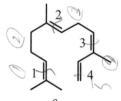

α-farnesene

359. How many moles of H₂ would be required to produce a saturated alkane from α-farnesene?

A. 1
B. 4
C. 8
D. 16

360. What reaction conditions could be used to produce acetone from α-farnesene?

A. H₅SO₄ and heat
B. HBr
C. O₃ and (CH₃)₂S
D. dilute acid and cold conditions

361. If α-farnesene was oxidized via ozonolysis such that all the double bonds reacted, how many products would result?

A. 2 products
B. 3 products
C. 4 products
D. 5 products

362. If α-farnesene is reacted with an excess of HBr, what would be the product?

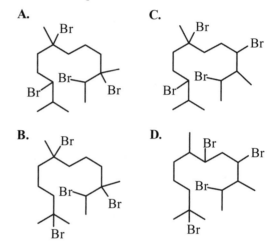

363. In the reaction of α-farnesene with excess HBr, which double bond would be the slowest to react?

A. The bond labeled 1.
B. The bond labeled 2.
C. The bond labeled 3.
D. The bond labeled 4.

364. What would be the product if α-farnesene was completely hydrated?

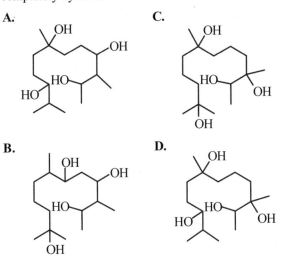

A. **C.** **B.** **D.**

365. The hydration of α-farnesene is an example of a(n):

A. radical reaction
B. combustion reaction
C. electrophilic addition
D. aromatic substitution

366. Why is the following reaction impractical to use on an industrial scale to produce the given product?

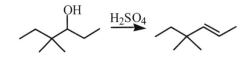

A. The carbocation intermediate leads to multiple products.
B. Dehydration of alcohols is not acid catalyzed.
C. Alkenes cannot be formed from an alcohol.
D. The alcohol is not available in large quantities.

367. Which of the following structures is a possible product of the hydrogenation of the alkene shown below using a Ni catalyst?

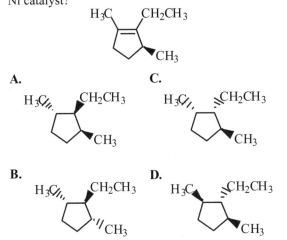

A. **C.** **B.** **D.**

368. How many possible alkane products are there for the hydrogenation of the alkene shown in the questions 367?

A. 1
B. 2
C. 3
D. 4

369. Dehydrohalogenation undergoes an E1 mechanism similar to dehydration. What are the possible products if 3-bromohexane undergoes dehydrohalogenation to form an alkene?

I.   *cis*-3-hexnene
II. *trans*-3-hexene
III. *cis*-2-hexene

A. I only
B. II only
C. I and III only
D. I, II, and III

370. When HBr is reacted with an alkene in the presence of a peroxide, the anti-Markovnikov product results. What would be the product if 2-isopropyl-2-butene is reacted with HBr in the presence of hydrogen peroxide?

A. 2-bromo-3-isopropylbutane
B. 2-bromo-2-isopropylbutane
C. 2-bromo-2-isopropylbutene
D. 2-isopropylbutane

371. Given that hydrogen peroxide is required to initiate the reaction and that the product is anit-Markovnikov, the reaction in question 370 is a(n):

A. radical reaction
B. combustion reaction
C. electrophilic addition
D. aromatic subsitituion

372. Hydroboration and oxidation of an alkene result in an anti-Markovnikov addition of a hydroxyl group. What is the product of the hydroboration and oxidation of 3,4-dimethyl-3pentene?

A. 3,4-dimethyl-2-pentanol
B. 3-isopropyl-2-butanol
C. 2,3-dimethyl-4-pentanol
D. 2-isopropylbutane

373. The reaction of vinyl chloride to form poly (vinyl chloride), PVC, is shown below. What type of reaction is this polymerization?

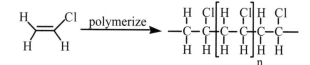

A. radical reaction
B. combustion reaction
C. electrophilic addition
D. aromatic substitution

374. What is the product when Cl_2 reacts with cyclopentene?

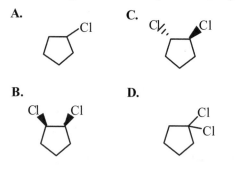

375. What is the product when Br_2 reacts with cyclohexene?

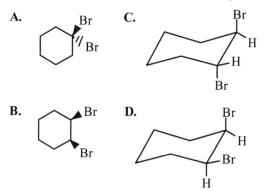

376. Which product would **not** form when 4-methyl-2-pentene is reacted with HBr?

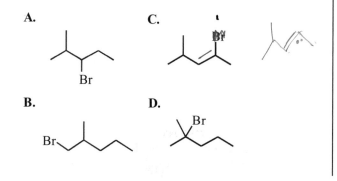

377. Which of the following is (are) possible products when 2-butene is reacted with HCl?

I. 2R-chlorobutane
II. 2S-chlorobutane
III. 2R, 2S-dichlorobutane

A. I only
B. II only
C. I and II only
D. I, II, and III

The results of the reaction of 2-pentene with potassium ethoxide and ethanol are shown below. Refer to the reaction to answer question 378-380.

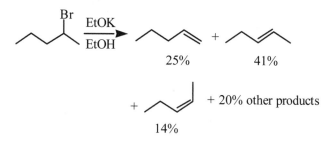

378. Why is 55% of the product 2-pentene compared to 25% of the product being 1-pentene?

A. 2-pentene is more reactive than 1-pentene.
B. 1-pentene is more reactive than 2-pentene.
C. 2-pentene is more thermodynamically stable than 1-pentene.
D. 1-pentene is more thermodynamically stable than 2-pentene.

379. Why is there approximately 3 times more *trans*-2-pentene than *cis*-2-pentene?

A. There is more steric hindrance in the trans isomer.
B. There is more steric hindrance in the cis isomer.
C. The cis isomer is the more thermodynamically stable isomer.
D. The reaction goes by a E2 mechanism.

380. The formation of pentenes from 2-bromopentane is a(n)

A. hydration
B. dehydration
C. dehydrohalogenation
D. hydrogenation

The reaction of 1,3-butadiene with HBr is shown below. At 40°C the major product is the 1,4-addition product; however, at –80°C the major product is the 1,2-addition product. Use this information and the reaction below to answer questions 381-386.

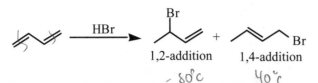

1,2-addition   1,4-addition

– 80°C   *40°C*

381. Which product is more thermodynamically stable?

  A. 1,2-addition product
  B. 1,4-addition product
  C. The products have the same stability.
  D. The relative stability cannot be determined.

382. Why are two products formed?

  A. There are two double bonds present.
  B. The carbocation intermediate allows delocalization of the second double bond.
  C. The fact that the carbocation is planar allows attack from both sides of the plane.
  D. There are 2 moles of HBr.

383. Which of the two products has a lower activation energy for formation?

  A. 1,2-addition product
  B. 1,4-addition product
  C. The products have the same activation energy.
  D. The relative activation energy cannot be determined.

384. How many products that are stereoisomers would form in the reaction of 1,3-butadiene with one mole of Br_2?

  A. 0
  B. 1
  C. 2
  D. 4

385. How many products that are structural isomers would form in the acid hydration of 1,3-butadiene?

  A. 0
  B. 1
  C. 2
  D. 4

386. How many moles of acetone form in the ozonolysis of one mole of 1,3-butadiene?

  A. 0
  B. 1
  C. 2
  D. 4

Friedel-Crafts acylation is a reaction of an acid chloride with an activated aryl ring. Use the example reaction shown below to answer questions 387-390.

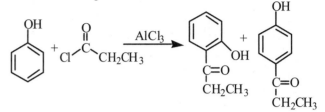

387. The Friedel-Crafts acylation is a(n):

  A. substitution reaction
  B. elimination reaction
  C. addition reaction
  D. redox reaction

388. Which of the following aryl compounds would be most reactive with an acid chloride to form an aryl ketone?

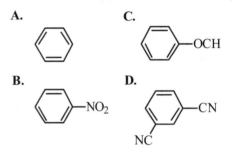

389. Which of the following aryl compounds would give only one aryl ketone when reacted with an acid chloride?

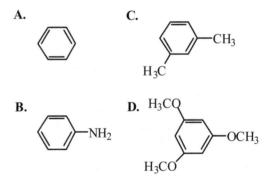

390. When reacted in excess acid chloride, toulene only produces one ketyl group on each aryl ring. What is the best explanation for this observation?

  A. steric hinderance
  B. The ketyl group is ring activating.
  C. The ketyl group is ring deactivating.
  D. resonance stabilization

Methyl benzoate is converted to methyl nitrobenzoate in the presence of HNO_3 and H_2SO_4. Use the reaction shown below to answer questions 391 - 393.

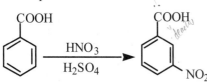

391. The acid group on the benzene ring is

A. meta directing and deactivating
B. ortho, para directing and deactivating
C. ortho, para directing and activating
D. meta directing and activating

392. Which of the following would be the easiest to nitrate?

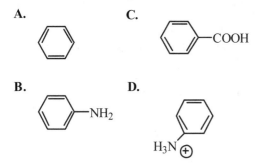

393. If phenol was reacted with HNO_3 in the place of benzoic acid, how many products would result?

A. 0
B. 1
C. 2
D. 3

Alcohols and Substitutions

394. In a S_N1 reaction, what are the bond conversions?

A. 2 sigma bonds are converted to 1 pi bond.
B. 1 sigma bond is converted into 2 pi bonds.
C. 1 pi bond is exchanged for 1 sigma bond.
D. 1 sigma bond is exchanged for 1 sigma bond.

395. In an S_N2 reaction, what are the bond conversions?

A. 2 sigma bonds are converted to 1 pi bond.
B. 1 sigma bond is converted into 2 pi bonds.
C. 1 pi bond is exchanged for 1 sigma bond.
D. 1 sigma bond is exchanged for 1 sigma bond.

396. The "S" in S_N2 is an abbreviation for what?

A. single step reaction
B. substitution
C. suspension
D. separation

397. What is true about a S_N1 reaction?

I. A carbocation intermediate is formed.
II. The rate determining step is bimolecular.
III. The mechanism has two steps.

A. I only
B. II only
C. I and III only
D. I, II, and III

398. What is true about an S_N2 reaction?

I. A carbocation intermediate is formed.
II. The rate determing step is bimolecular.
III. The mechanism has two steps.

A. I only
B. II only
C. I and III only
D. I, II, and III

399. The rate of a S_N2 reaction depends on:

A. the concentration of the nucleophile only.
B. the concentration of the electrophile only.
C. the concentration of both the nucleophile and the electrophile.
D. neither the concentration of the nucleophile nor the electrophile.

400. The rate of a S_N1 reaction depends on:

A. the concentration of the nucleophile only.
B. the concentration of the electrophile only.
C. the concentration of both the nucleophile and the electrophile.
D. neither the concentration of the nucleophile nor the electrophile.

401. How does the rate of step I in the mechanism shown below compare to the rate of step II?

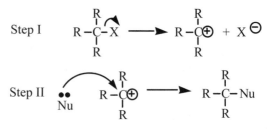

A. Step I is faster than step II.
B. Step II is faster than step I.
C. Step I and step II happen at the same rate.
D. The relationship cannot be determined.

402. Which of the following is the best nucleophile?

- **A.** CO_2
- **B.** H_2O
- **C.** CH_3O^-
- **D.** CH_3OH

403. Which of the following is the best nucleophile?

- **A.** I^-
- **B.** H_2O
- **C.** NH_4^+
- **D.** CH_3COOH

404. Which of the following is the best leaving group?

- **A.** I^-
- **B.** OH^-
- **C.** NH_4^+
- **D.** F^-

405. Which of the following is the best leaving group?

- **A.** H_2O
- **B.** OH^-
- **C.** NH_4^+
- **D.** CH_3COO^-

406. A nucleophile is:

- **A.** a Lewis acid.
- **B.** a Lewis base.
- **C.** electron deficient .
- **D.** an oxidizing agent.

407. Which of the following alkyl halogens would react the fastest in a reaction with sodium hydroxide?

- **A.** t-butyl fluoride
- **B.** t-butyl chloride
- **C.** t-butyl bromide
- **D.** t-butyl iodide

408. Which of the following reactions is most likely to proceed by a S_N2 mechanism?

- **A.** 1-bromopropane with sodium hydroxide
- **B.** 2-bromo-2-methyl pentane with HBr
- **C.** 2-bromo-3-methyl pentane with methanol
- **D.** t-butyl iodide with ethanol

409. Which of the following reactions is most likely to proceed by a S_N1 mechanism?

- **A.** 1-bromopropane with sodium hydroxide
- **B.** 2-bromo-4-methyl pentane with HBr
- **C.** 1-bromo-3-methyl pentane with HCl
- **D.** t-butyl iodide with ethanol

410. Which of the following alcohols would have the lowest boiling point?

- **A.** ethanol
- **B.** 1-butanol
- **C.** 2-methyl-1-propanol
- **D.** 1-pentanol

411. Which of the following alcohols would have the highest boiling point?

- **A.** ethanol
- **B.** 1-butanol
- **C.** 2-methyl-1-propanol
- **D.** 1-pentanol

412. Given that the boiling point for 1-propanol is 97°C and the boiling point of 1-pentanol is 138°C, what is the boiling point of 1-butanol?

- **A.** 100°C
- **B.** 118°C
- **C.** 138°C
- **D.** 150°C

413. Given that the boiling point for 3-methyl-1-butanol is 132°C and the boiling point of 1-pentanol is 138°C, what is the boiling point of 1-hexanol?

- **A.** 100°C
- **B.** 132°C
- **C.** 138°C
- **D.** 156°C

414. Which of the following would have the highest boiling point?

- **A.** 1-butanol
- **B.** butane
- **C.** 1-butene
- **D.** 1-butyne

415. Given that heptane has a boiling point of 98°C, what phase will heptanol be in at 98°C?

- **A.** gas
- **B.** liquid
- **C.** solid
- **D.** supercritical fluid

416. Which of the following has the highest boiling point?

- **A.** dimethyl ether
- **B.** diethyl ether
- **C.** ethyl methyl ether
- **D.** diisopropyl ether

417. Which of the following has the highest boiling point?

    A. ethane
    B. diethyl ether
    C. ethene
    D. ethanol

418. Cyclohexanol is a solid in a cool room, while cyclohexane is a liquid, what is the best explanation for the different states?

    A. Cyclohexanol has greater London Dispersion forces than cyclohexane.
    B. Cyclohexanol has weaker London Dispersion forces than cyclohexane.
    C. Cyclohexanol has hydrogen bonding and cyclohexane does not.
    D. Cyclohexane has hydrogen bonding and cyclohexanol does not.

419. Dimethyl ether has a boiling point of -25°C and propane has a boiling point of -42°C. Why does dimethyl ether have a greater boiling point?

    A. Dimethyl ether has hydrogen bonding and propane does not.
    B. Propane has a greater molecular weight than dimethyl ether.
    C. Dimethyl ether has a smaller dipole moment than propane.
    D. Dimethyl ether has a larger dipole moment than propane.

420. Which of the following would be more soluble in water?

    A. 1-butanol
    B. butane
    C. 1-butene
    D. 1-butyne

421. Which of the following would be more soluble in water?

    A. diethyl ether
    B. ethene
    C. ethyne
    D. ethane

422. The word miscible means that one solvent forms a homogenous solution when mixed with another solvent in any amount. Which of the following would be miscible with water?

    A. ethanol
    B. cyclohexane
    C. 2-butene
    D. hexane

423. Which of the following would be miscible with water?

    A. 1-propanol
    B. phenol
    C. 4-methyl-1-octanol
    D. 2-methyl-2-propanol

424. Which of the following will have the most acidic proton?

    A. pentane
    B. 3-pentene
    C. 1-pentyne
    D. 1-pentanol

425. What is 2-methyl-2-butanol?

    A. primary alcohol
    B. secondary alcohol
    C. tertiary alcohol
    D. quaternary alcohol

426. Which of the following will have the most acidic proton?

    A. phenol
    B. ethanol
    C. t-butyl alcohol
    D. 2-methyl-1-hexanol

427. Grignard reactions require a solvent that is polar but that does not contain an acidic proton. Which of the following would be the best solvent?

    A. hexane
    B. ethanol
    C. acetic acid
    D. diethyl ether

428. In the following reaction of ethanol with para-toluenesulfonylchloride, ethanol is a(n):

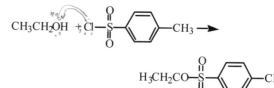

    A. nucleophile
    B. electrophile
    C. radical
    D. reducing agent

429. The reaction of ethanol with para-toluenesulfonylchloride, shown in question 428, is a:

    A. nucleophilic substitution
    B. nucleophilic addition
    C. reductive elimination
    D. syn addition

430. Which of the following will convert a secondary alcohol to ketone?

   A. $LiAlH_4$
   B. PBr_3
   C. K_2CrO_4
   D. O_3 and $(CH_3)_2S$

431. Which of the following will convert a primary alcohol to an aldehyde?

   A. $LiAlH_4$
   B. dilute cold $KMnO_4$
   C. K_2CrO_4
   D. O_3 and $(CH_3)_2S$

432. Which of the following will convert a primary alcohol to an carboxylic acid?

   A. $LiAlH_4$
   B. dilute cold $KMnO_4$
   C. K_2CrO_4
   D. O_3 and $(CH_3)_2S$

433. The reaction of 3-methyl-(S)-3-hexanol with HBr produces an optically inactive solution. What is the best explanation for this result?

   A. The product does not have a chiral carbon.
   B. The product is meso.
   C. The reaction produces a racemic mixture of products.
   D. All alkyl halide solutions are optically inactive.

The Lucas test is used to determine whether an alcohol is primary, secondary, or tertiary based on the rate of reaction with HCl and $ZnCl_2$. Use the table below to answer questions 434 - 442.

Class of alcohol	Approximate time of reaction.	Mechanism
Primary	> 6 minutes	S_N2
Secondary	1-5 minutes	S_N1
Tertiary	< 1 minutes	S_N1

434. A scientist wanted to increase the initial rate of reaction of t-butyl alcohol with HCl and $ZnCl_2$. Which of the following will speed up this reaction?

   A. Increase the concentration of t-butyl alcohol
   B. Increase the concentration of HCl
   C. Increase the concentration of $ZnCl_2$
   D. Decrease the temperature of the reaction

435. Which of the following alcohols will reacted the slowest?

   A. 1-pentanol.
   B. 2-methyl-3-pentanol
   C. 2-methyl-2-pentanol
   D. 2-hexanol

436. Why do tertiary alcohols react faster than secondary alcohols with HCl and $ZnCl_2$?

   A. Secondary alcohols are less hindered.
   B. Tertiary alcohols are less hindered
   C. Secondary alcohols form more stable carbocations.
   D. Tertiary alcohols form more stable carbocations.

437. What is the purpose of the $ZnCl_2$ in the Lucas reagent?

   A. To activate the alcohol and make it a better leaving group
   B. To provide a source of chloride ions to complete the substitution
   C. To activate the chloride ion from HCl and make it a better nucleophile
   D. To activate the chloride ion from HCl and make it a better electrophile

438. Which of the following would be true of a reaction of 2-ethyl-(2R)-pentanol with the Lucas reagent?

   A. The product would be optically active.
   B. Increasing the concentration of HCl would increase the rate of reaction.
   C. 2-ethyl-(2R)-pentanol would react slower than pentanol.
   D. A carbocation is formed during the reaction.

439. What would be the product of a reaction of 1-propanol with the Lucas reagent?

   A. 1-chloropropane
   B. 2-chloro-1-propanol
   C. 1-propene
   D. propane

440. An unknown alcohol is reacted with the Lucas reagent and the reaction is completed in 45 seconds. The unknown alcohol is a(n):

   A. tertiary alcohol
   B. secondary alcohol
   C. primary alcohol
   D. The type of alcohol cannot be determined.

441. Which of the following would be true of a reaction of the unknown alcohol in question 440 with the Lucas reagent?

   A. The product would be optically active.
   B. Increasing the concentration of HCl would increase the rate of reaction.
   C. Increasing the concentration of the unknown alcohol would increase the rate of reaction.
   D. Propanol would react at a faster rate with the Lucas reagent.

442. An unknown alcohol is reacted with the Lucas reagent and the reaction is complete in 8 minutes. The unknown alcohol is

   A. 1-butanol
   B. 2-pentanol
   C. 3-methyl-3-hexanol
   D. cyclohexanol

443. Which of the following reagents will convert 1-butanol to 1-bromobutane?

   A. PBr_3
   B. Br_2
   C. $LiAlH_4$
   D. $KMnO_4$

444. Which of the following reagents will convert diethyl ether to ethanol and chloroethane?

   A. HCl
   B. Cl_2
   C. $SOCl_2$
   D. $AlCl_3$

445. Which of the following reagents will convert 3-hexanone to 3-hexanol?

   A. $NaBH_4$
   B. H_2CrO_4
   C. $KMnO_4$
   D. $AlCl_3$

446. Why does the presence of an acid facilitate the substitution of an alcohol?

   A. The acid converts the OH group to a good nucleophile.
   B. The acid converts the OH group to a good electrophile.
   C. The acid converts the OH group to a good leaving group.
   D. The acid neutralizes the bases in solution.

447. What is the product of the following reaction?

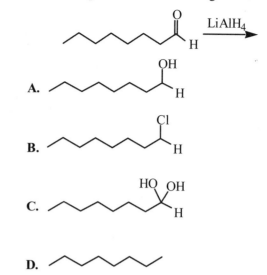

   A.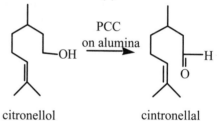

   B.

   C.

   D.

448. Citronellol is converted to citronellal by PCC (pyridinium chlorochromate) according the reaction shown below. PCC is a(n):

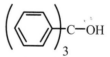

   citronellol   cintronellal

   A. oxidizing agent
   B. reducing agent
   C. acid
   D. base

449. Which of the following could be used to convert the cintronellal back to cintronellol (question 448)?

   A. PBr_3
   B. Br_2
   C. $LiAlH_4$
   D. $KMnO_4$

Refer to the structure of triphenylmethanol shown below to answer questions 450 - 454.

450. Which of the following reagents would form triphenylmethyl bromide?

   A. HBr in dilute acetic acid
   B. HBr in sodium hydroxide
   C. Br_2
   D. CH_3Br in methanol

451. How would the rate of triphenylmethyl alcohol compare to the rate of *t*-butyl alcohol if the same reaction conditions used in the bromination reaction described in question 450 were used?

   A. t-Butyl alcohol would have a faster rate.
   B. Triphenylmethyl alcohol would have a faster rate.
   C. The rates would be the same.
   D. The relationship of the rates cannot be determined.

452. By what mechanism is triphenylmethyl bromide formed in the reaction described in question 450?

   A. S_N2
   B. S_N1
   C. E1
   D. E2

453. When triphenylmethanol is dissolved in concentrated sulfuric acid and transferred to a cold solution of methanol, what is the structure of the solid that forms?

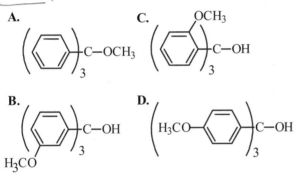

454. In the reaction in question 453, methanol is a(n)

   A. electophile
   B. nucleophile
   C. reducing agent
   D. ether

The structure of testosterone, the male sex hormone, is shown below. Refer to the structure to answer questions 455-457.

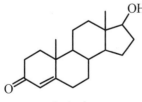

testosterone

455. Which of the reagents will give the product shown below when reacted with testosterone?

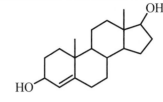

   A. $LiAlH_4$
   B. PBr_3
   C. K_2CrO_4
   D. O_3 and $(CH_3)_2S$

456. Which of the reagents will give the product shown below when reacted with testosterone?

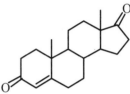

   A. $LiAlH_4$
   B. PBr_3
   C. K_2CrO_4
   D. O_3 and $(CH_3)_2S$

457. Why is HBr a poor choice as a reagent to do an electrophilic addition of bromine to the double bond in testosterone?

   A. The bromine placement on the double bond would not be selective for a specific carbon in the double bond.
   B. The HBr would also promote a substitution at the alcohol position.
   C. HBr does not react with alkenes.
   D. The reaction would be too slow.

458. When 1-chlorobutane and 1-bromobutane are reacted with sodium iodide in acetone, the appearance of a precipitate indicates the completion of the reaction. What is the best explanation for the fact that precipitate appears in the reaction of 1-bromobutane first?

   A. Cl⁻ is a better leaving group than Br⁻.
   B. Br⁻ is a better leaving group than Cl⁻.
   C. Cl⁻ is a better nucleophile than Br⁻.
   D. Br⁻ is a better nucleophile than Cl⁻.

459. By what mechanism do the reactions described in question 458 proceed?

   A. S_N2
   B. S_N1
   C. E1
   D. E2

460. The rate of precipitate formation increases for the reaction of 1-bromobutane with NaI as the concentration of NaI is increased. When the concentration of NaI is increased in the reaction of 2-chloro-2-methylpropane, no rate change is observed. Why is there a change in rate in the first case and not in the second?

A. Cl⁻ is a better leaving group than Br⁻.
B. Br⁻ is a better leaving group than Cl⁻.
C. 1-bromobutane proceeds via a S_N2 mechanism, while 2-chloro-2-methylpropane proceeds via a S_N1 mechanism.
D. 1-bromobutane proceeds via a S_N1 mechanism, while 2-chloro-2-methylpropane proceeds via a S_N2 mechanism.

461. What would be the product when NaBr is reacted (S)-1-iodo-1-ethanol?

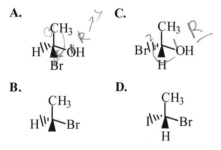

462. $KMnO_4$ forms a syn diol when reacted in a cold and dilute solution with an alkene. Oxidative cleavage results when it is reacted in a warm and concentrated solution. What would be the product if cold dilute $KMnO_4$ is reacted with 3-cyclohexene-1-ol?

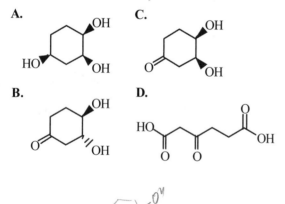

463. Given the information in question 462, what would be the product if warm concentrated $KMnO_4$ is reacted with 3-cyclohexene-1-ol?

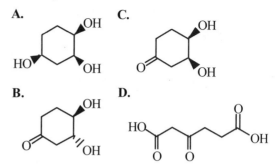

464. When (2R)-bromobutane is reacted with NaI the product is optically active. What is the mechanism for the reaction?

A. S_N2
B. S_N1
C. E1
D. E2

465. Dehydration of 2-hexanol to form 2-hexene and 1-hexene proceeds by what mechanism?

A. S_N2
B. S_N1
C. E1
D. E2

466. In a portable breath tester, fuel cells catalyze the oxidation of ethanol by oxygen. What is the product of this reaction?

A. ethanol
B. 1-hydroxy-ethanol
C. 2-hydroxy-ethanol
D. ethanal

467. Alcohols dry human skin as they evaporate from the skin's surface. Why does ethanol have a greater drying effect than isopropanol?

A. Ethanol has a higher boiling point than isopropanol.
B. Ethanol has a lower boiling point than isopropanol.
C. Ethanol hydrogen bonds more strongly with skin than isopropanol.
D. Isopropanol hydrogen bonds more strongly with skin than ethanol.

468. Equal amounts of ammonium chloride (0.25 moles) and ammonium bromide (0.25 moles) are dissolved in an acidic solution. After the salts have dissolved 0.20 moles of 1-pentanol is added to the solution. What is the product of this reaction?

   A. 1-pentanol
   B. 1-bromopentane
   C. 1-chloropentane
   D. equal amounts of 1-bromopentane and 1-chloropentane

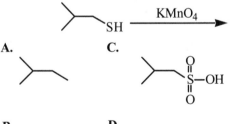

469. Why must the solution be acidic in the reaction described in question 468?

   A. The acid makes the OH group into a better leaving group.
   B. The acid makes the OH group into a better nucleophile.
   C. The acid makes Br or Cl into better leaving groups.
   D. The acid makes Br or Cl into better nucleophiles.

470. By what mechanism do the reactions described in question 468 proceed?

   A. S_N2
   B. S_N1
   C. E1
   D. E2

471. Methyl mercaptan (CH_3SH) is similar to:

   A. a primary alcohol.
   B. a secondary alcohol.
   C. a tertiary alcohol.
   D. a quaternary alcohol.

472. What would be the product of the reaction shown below?

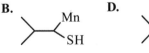

The reaction of (2R)-iodobutane to form 2-methoxybutane proceeds by an S_N1 mechanism in methanol. However, in the presence of the methoxide ion (CH_3O^-) the mechanism is S_N2. Use this information to answer questions 473 – 479.

473. Why does the mechanism change when methanol is replaced by methoxide?

   A. Iodide is a poor leaving group.
   B. Iodide is a good leaving group.
   C. Methanol is a weak nucleophile and methoxide is a strong nucleophile.
   D. Methanol is a strong nucleophile and methoxide is a weak nucleophile.

474. Given the above information, what would be the product of the reaction of (2R)-iodobutane with methanol?

   I. (2R)-methoxybutane
   II. (2S)-methoxybutane
   III. butane

   A. I only
   B. II only
   C. I and II only
   D. I, II, and III

475. Given the above information, what would be the product of the reaction of (2R)-iodobutane with the methoxide ion?

   I. (2R)-methoxybutane
   II. (2S)-methoxybutane
   III. butane

   A. I only
   B. II only
   C. I and II only
   D. I, II, and III

476. Which of the following alkanes would react faster than (2R)-iodobutane with methanol to form 2-methoxybutane?

   A. 1-iodobutane
   B. 2-bromobutane
   C. 2-iodopropane
   D. 2-iodo-2-methylbutane

477. Which of the following alkanes would react faster than (2R)-iodobutane with the methoxide ion to form 2-methoxybutane?

   A. 1-iodobutane
   B. 2-bromobutane
   C. 2-iodopropane
   D. 2-iodo-2-methylbutane

478. Which of the following statements regarding the reaction of (2R)-iodobutane with methanol is NOT true?

    A. The reaction proceeds with complete inversion of the stereochemistry.
    B. The product solution will not be optically active.
    C. Increasing the concentration of (2R)-iodobutane will increase the reaction rate.
    D. Iodide is a good leaving group.

479. Which of the following statements regarding the reaction of (2R)-iodobutane with the methoxide ion is NOT true?

    A. The reaction proceeds with complete inversion of the stereochemistry.
    B. The product solution will not be optically active.
    C. Increasing the concentration of (2R)-iodobutane will increase the reaction rate.
    D. Iodide is a good leaving group.

480. The reaction of 1-chlorobutane with a boiling sodium hydroxide solution is a two phase reaction. When the two phase reaction is complete, why does a homogenous solution result?

    A. The reaction is exothermic.
    B. The second solution is boiled off.
    C. The haloalkane product is soluble in water.
    D. The alcohol product is soluble in water.

481. Methanol is made commercially by the hydrogenation of carbon monoxide. The reaction is a(n):

    A. oxidation
    B. reduction
    C. elimination
    D. substitution

Shown below is the structure of sodium cholate which is a detergent used in the purification of proteins. Refer to the structure to answer questions 482 - 485.

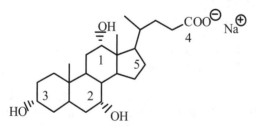

482. Which of the three hydroxyl carbons would be most likely to participate in a S$_N$2 reaction?

    A. The carbon labeled 1.
    B. The carbon labeled 2.
    C. The carbon labeled 3.
    D. All three would have the same reactivity.

483. Which of the following reagents would convert sodium cholate to the product shown below?

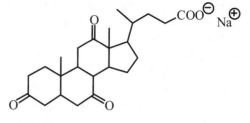

    A. LiAlH$_4$
    B. PBr$_3$
    C. K$_2$CrO$_4$
    D. O$_3$ and (CH$_3$)$_2$S

484. Which of the carbons in sodium cholate has the highest oxidation state?

    A. The carbon labeled 1
    B. The carbon labeled 3
    C. The carbon labeled 4
    D. The carbon labeled 5

485. Which of the following statements is NOT true about sodium cholate?

    A. Sodium cholate is optically active.
    B. The hydroxy groups can be oxidized to aldehydes.
    C. Sodium cholate is soluble in water.
    D. Sodium cholate is not a planar molecule.

486. What is the product when glyceraldehyde (shown below) is reacted with an excess of K$_2$CrO$_4$?

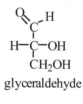
glyceraldehyde

    A.

    C.

    B.

    D.

Refer to the scheme below to answer questions 487 – 496.

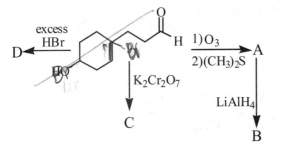

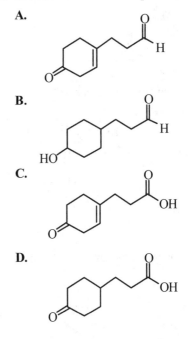

487. How does the oxidation state of product B compare to the oxidation state of product A?

A. The oxidation state of A is higher than B.
B. The oxidation state of B is higher than A.
C. The two products have the same oxidation state.
D. The oxidation state of the two products cannot be compared.

488. When K_2CrO_7 is added to the starting material shown in the scheme, the K_2CrO_7

A. gains electrons.
B. loses electrons.
C. maintains the same number of electrons.
D. cannot transfer electrons.

489. How many carbonyl groups does product A have?

A. 1
B. 2
C. 3
D. 4

490. How many hydroxyl groups does product B have?

A. 1
B. 2
C. 3
D. 4

491. In the reaction of HBr with the hydroxyl group of the starting material, the Br^- is a(n)

A. electrophile.
B. nucleophile.
C. acid.
D. reducing agent.

492. In the reaction of HBr with the hydroxyl group of the starting material, what is the first step of the mechanism?

A. attack of the carbon by Br^-
B. formation of a carbocation by the loss of ^-OH
C. inversion of stereochemistry
D. protonation of the OH group

493. How many chiral centers does product D have?

A. 0
B. 1
C. 2
D. 3

494. The correct structure for product C is

A.

B.

C.

D.

495. When HBr adds to the starting alkene, the addition is:

A. a Markovnikov eletrophilic addition.
B. an anti-Markovnikov eletrophilic addition.
C. a Markovnikov nucleophilic addition.
D. an anti-Markovnikov nucleophilic addition.

496. What reagent could be used to convert product C to a diol?

A. $LiAlH_4$
B. PBr_3
C. K_2CrO_4
D. O_3 and $(CH_3)_2S$

Aldehydes and Ketones

497. Which of the following structures is an aldehyde?

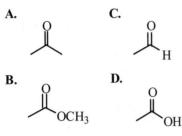

A.

C.

B.

D.

498. Which of the following structures is a ketal?

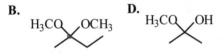

A.

C.

B.

D.

499. Which of the following structures is a hemiacetal?

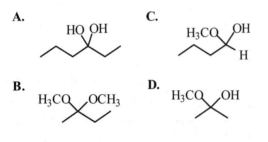

A.

C.

B.

D.

500. Which of the following structures is a hemiketal?

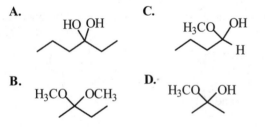

A.

C.

B.

D.

501. Which of the following is NOT true of carbonyl compounds?

A. The carbonyl carbon has a partial positive charge.
B. The carbonyl oxygen has a partial negative charge.
C. The carbonyl carbon is subject to electrophilic attack.
D. The carbonyl group has planar stereochemistry.

502. In the structure shown below, which atoms are in the same plane?

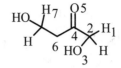

A. Atoms H1, C4, and O3.
B. Atoms H7, H1, and C4.
C. Atoms C2, C4, and C6.
D. Atoms O5, O3, and H1.

503. Which of the following compounds is the most acidic?

A. pentane
B. 1-pentene
C. 2-pentanone
D. pentanal

504. In the structure shown below, which hydrogen is the most acidic?

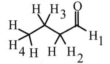

A. The hydrogen atom labeled 1.
B. The hydrogen atom labeled 2.
C. The hydrogen atom labeled 3.
D. The hydrogen atom labeled 4.

505. Which of the following hydrogen atoms is a beta hydrogen?

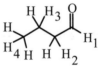

A. The hydrogen atom labeled 1.
B. The hydrogen atom labeled 2.
C. The hydrogen atom labeled 3.
D. The hydrogen atom labeled 4.

506. Which of the following hydrogen atoms is a gamma hydrogen?

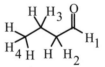

A. The hydrogen atom labeled 1.
B. The hydrogen atom labeled 2.
C. The hydrogen atom labeled 3.
D. The hydrogen atom labeled 4.

507. In the structure shown below, which hydrogen is the most acidic?

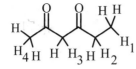

A. The hydrogen atom labeled 1.
B. The hydrogen atom labeled 2.
C. The hydrogen atom labeled 3.
D. The hydrogen atom labeled 4.

508. In the structure shown below, which carbon is the most electrophilic?

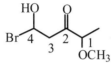

A. The carbon atom labeled 1.
B. The carbon atom labeled 2.
C. The carbon atom labeled 3.
D. The carbon atom labeled 4.

509. In the structure shown below, which carbon is most nucleophilic?

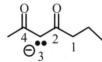

A. The carbon atom labeled 1.
B. The carbon atom labeled 2.
C. The carbon atom labeled 3.
D. The carbon atom labeled 4.

510. Which of the following is a resonance structure of 2-pentanone?

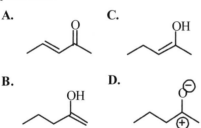

511. Why are the alpha hydrogens of ketones more acidic than the hydrogens of alkanes?

A. The oxygen of the ketone promotes hydrogen bonding.
B. The oxygen of the ketone donates electrons to the base.
C. The ketone can resonance stabilize the anions generated when the proton is removed.
D. The ketone can resonance destabilize the anions generated when the proton is removed.

512. Which of the following structures shows the correct net dipole moment for acetone?

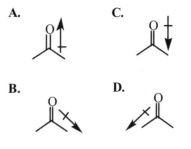

Refer to the structures below to answer questions 513 - 515.

Structure A   Structure B

513. What is the relationship between structures A and B?

A. geometric isomers
B. stereoisomers
C. tautomers
D. There is no relationship.

514. Structure A is a(n):

A. enol form
B. keto form
C. enatiomer
D. oxidant

515. Structure B is a(n):

A. enol form
B. keto form
C. enantiomer
D. oxidant

516. Which of the following will have the highest boiling point?

A. 1-heptanol
B. heptane
C. 2-hepanone
D. heptanal

517. Which of the following will have the greatest water solubility?

A. 1-heptanol
B. heptane
C. 2-hepanone
D. heptanal

518. Which of the following will have the highest boiling point?

   A. 2-propanone
   B. 2-butanone
   C. 2-pentanone
   D. 2-hexanone

519. Which of the following will have the highest water solubility?

   A. 2-propanone
   B. 2-butanone
   C. 2-pentanone
   D. 2-hexanone

520. If 2-pentanone has a water solubility of 5.5 % by volume, what would be true about a solution made with 5 mL of 2-butanone and 95 mL of water?

   A. Two layers would be observed with one layer having a significantly smaller volume.
   B. Two layers would be observed with both layers having the same volume.
   C. The solution would be homogenous.
   D. The activity of 2-butanone cannot be predicted based on 2-pentanone.

521. Which of the following accounts for the fact the acetone is miscible in water?

   I. Acetone has a small molecular weight.
   II. Acetone can form a hydrogen bond with water.
   III. Acetone molecules hydrogen bond with other acetone molecules.

   A. I only
   B. III only
   C. I and II only
   D. I and III only

522. Given that ethanal (acetaldehyde) has a boiling point of 21°C, in what phase is formaldehyde (methanal) at room temperature?

   A. gas
   B. liquid
   C. solid
   D. super critical liquid

523. In the reaction of an aldehyde with the Tollen's reagent, Ag^+ is reduced to Ag which produces a silver mirror. Why does the silver mirror not appear when Tollen's reagent is added to a ketone?

   A. Ketones have an acidic proton and aldehydes do not.
   B. Aldehydes have an acidic proton and ketones do not.
   C. Ketones can be oxidized while aldehydes cannot.
   D. Aldehydes can be oxidized while ketones cannot.

524. Given the information about the Tollen's reagent in questions 523, which of the following will react with the Tollen's reagent?

   A. pentanal
   B. 3-hexanone
   C. hexanoic acid
   D. 2-heptene

525. Given the information about the Tollen's reagent in questions 523, what are the products when butanal is reacted with the Tollen's reagent?

   A. butanoic acid and Ag^+
   B. butanoic acid and Ag
   C. 1-butanol and Ag^+
   D. 1-butanol and Ag

526. What reagents could be used to synthesize the acetal shown below?

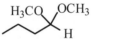

   A. 1-butanol with methanal
   B. butanal with methanol
   C. 1,1-dibutanol and water
   D. propanal with methanol

527. Butyraldehyde has a buttery odor and is used in margarine. What reagent could be used to synthesize butyraldehyde from 1-butanol?

   A. $K_2Cr_2O_7$
   B. PCC (mild oxidant)
   C. $LiAlH_4$
   D. O_3, Zn

Terminal alkynes can be used to synthesize methyl ketones. Refer to the example of this reaction below to answer questions 528 - 532.

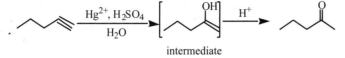

528. Step one in the formation of 2-pentanone from 1-pentyne is a(n):

A. hydration
B. dehydration
C. hydrogenation
D. nucleophilic substitution

529. Step two in the formation of 2-pentanone from 1-pentyne is a(n):

A. oxidation
B. reduction
C. tautomerization
D. nucleophilic substitution

530. The intermediate in the synthesis is a(n):

A. alcohol
B. enol
C. alkene
D. isolated product

531. Why does the reaction shown for the synthesis of 2-pentanone only form a methyl ketone and not an aldehyde?

A. Only the Markovnikov alcohol product forms.
B. Only the anti-Markovnikov alcohol product forms.
C. Aldehydes are more unstable than ketones.
D. Aldehydes are more acidic than ketones.

532. What would be the product if 1-pentene was used instead of 1-pentyne?

A. 1-pentanol
B. 2-pentanol
C. pentanal
D. 2-pentanone

When reacted with water, formaldehyde forms formalin, used to preserve biological specimens. Use the reaction shown below to answer questions 533 – 539.

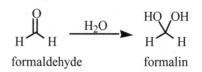

533. The formation of formalin from flormaldehyde is a(n):

A. electrophilic elimination
B. aromatic subsitituion
C. nucleophilic addition
D. nucleophilic substitution

534. In a base catalyzed formation of formalin from formaldehyde, the nucleophile is:

A. the carbonyl carbon in formaldehyde.
B. the carbonyl oxygen in formaldehyde.
C. OH^-
D. H_2O

535. In a base catalyzed formation of formalin from formaldehyde, the electrohpile is:

A. the carbonyl carbon in formaldehyde.
B. the carbonyl oxygen in formaldehyde.
C. OH^-
D. H_2O

536. In an acid catalyzed formation of formalin from formaldehyde, what does the acid catalyst do?

A. Activates the electrophile by protonating the oxygen of the carbonyl.
B. Activates the nucleophile by protonating the oxygen of the carbonyl.
C. Activates the electrophile by protonating the oxygen of water.
D. Activates the nucleophile by protonating the oxygen of water.

537. In an acid catalyzed formation of formalin from formaldehyde, the nucleophile is

A. the carbonyl carbon in formaldehyde.
B. the carbonyl oxygen in formaldehyde.
C. OH^-
D. H_2O

538. In the above reaction, the formation of the product (formalin) is favored over the starting compound (formaldehyde). However, when this reaction is conducted with 2 butanone as the reactant, the starting compound is favored over the hydrated product. What accounts for the difference in reactivity?

   A. Alkyl groups are electron donating.

   B. Alkyl groups are electron withdrawing.

   C. Formaldehyde is sterically hindered.

   D. Ketones are weaker nucleophiles.

539. What would result if formaldehyde was dissolved in methanol rather than water?

   A. The formation of hydrated formaldehyde

   B. The formation of an acetal

   C. The formation of a ketal

   D. No reaction would occur

540. What reagent could form the camphor product from the borneol starting material?

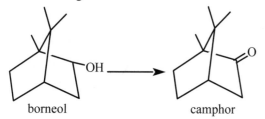

     borneol   camphor

   A. NaBH$_4$

   B. PBr$_3$

   C. Na$_2$Cr$_2$O$_7$

   D. O$_3$

541. What reagent could convert camphor to borneol (see question 540)?

   A. NaBH$_4$

   B. PBr$_3$

   C. Na$_2$Cr$_2$O$_7$

   D. O$_3$

542. What nucleophile could be used to form cyanohydrin (shown below) when reacted with 2-hexanone?

     cyanohydrin

   A. HCN

   B. CN⁻

   C. H$_2$O

   D. OH⁻

Use the generic Grignard reaction shown below to answer questions 543 - 547.

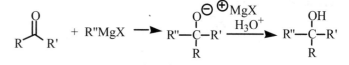

543. In a Grignard reaction the Grignard reagent (R"MgX) is a:

   A. strong nucleophile

   B. weak nucleophile

   C. strong electrophile

   D. weak electrophile

544. Which of the following solvents would be best for a Grignard reaction?

   A. water

   B. ethanol

   C. diethyl ether

   D. dichloromethane

545. If 2-pentanone is reacted with phenyl magnesium bromide what is true about the product?

   A. The product has a chiral center and the solution is optically active.

   B. The product has a chiral center and the solution is not optically active.

   C. The product has no chiral center and the solution is not optically active.

   D. The product is a meso compound.

546. The Gringnard reaction is a(n):

   A. electrophilic elimination

   B. aromatic subsitituion

   C. nucleophilic addition

   D. nucleophilic substitution

547. Which of the following is a Grignard reagent?

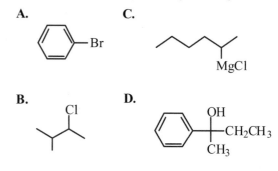

548. In an aldol addition of 2-butanone and 3-pentanone the product will have:

- **A.** 2 carbonyl groups.
- **B.** 2 hydroxyl groups.
- **C.** 1 carbonyl group, 1 hydroxyl group.
- **D.** 1 carbonyl group, 1 hydroxyl group, and one carbon carbon double bond.

549. In an aldol reaction the nucleophile is a(n):

- **A.** carbonyl carbon
- **B.** carbonyl oxygen
- **C.** hydroxide ion
- **D.** enolate ion

550. What is the first step of a base catalyzed aldol condensation?

- **A.** Protonation of the carbonyl oxygen to activate the nucleophile
- **B.** Protonation of the enolate ion to activate the nucleophile
- **C.** Deprotonation of the alpha hydrogen to activate the nucleophile
- **D.** Deprotonation of the alpha hydrogen to activate the electrophile

551. How does a base catalyzed aldol condensation compare with an acid catalyzed aldol condesation?

- **A.** An acid catalyzed aldol condensation activates the electrophile, while the base catalyzed condensation activates the nucleophile.
- **B.** An acid catalyzed aldol condensation activates the nucleophile, while the base catalyzed condensation activates the electrophile.
- **C.** Both reactions activate the electrophile.
- **D.** Both reactions activate the nucleophile.

552. What must be done to an aldol product to obtain an α, β-unsaturated carbonyl?

- **A.** hydration
- **B.** dehydration
- **C.** hydrogenation
- **D.** oxidation

553. A crossed aldol reaction between ethanal and propanal gives how many products that are structural isomers?

- **A.** 1
- **B.** 2
- **C.** 3
- **D.** 4

554. Which of the following is a product of the aldol condensation of 2-butanone?

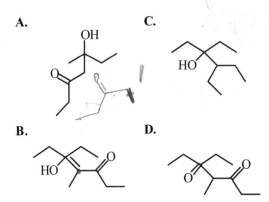

A.   **C.**

B.   **D.**

555. If a scientist desired only one product from a crossed aldol reaction, which of the following aldehydes should be reacted in excess with 2-propone?

A.   **C.**

B.   **D.**

Use the reaction shown below to answer questions 556-558.

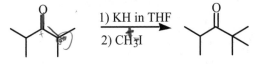

$$\text{1) KH in THF}$$
$$\text{2) CH}_3\text{I}$$

556. Why does the methyl group add to the alpha carbon rather than the carbonyl carbon in the reaction shown above?

- **A.** The alpha carbon is acidic.
- **B.** The beta carbon is acidic.
- **C.** The carbonyl is not electrophilic.
- **D.** The KH forms the enolate ion.

557. What type of reaction is the addition of the methyl group to form 2,2,4-trimethyl-3-pentanone?

- **A.** electrophilic addition
- **B.** nucleophilic subsitution
- **C.** conjugate addition
- **D.** elimination

558. In the addition of the methyl group to form 2,2,4-trimethyl-3-pentanone, the methyl iodide is a(n):

A.   electrophile
B.   nucleophile
C.   oxidant
D.   reductant

559. If a strong nucleophile is added to the compound shown below, which carbons are possible electrophiles?

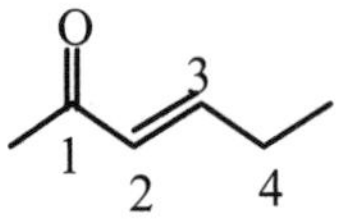

A.   only C_1
B. C_1 and C_2
C. C_1 and C_3
D. C_2 and C_2

560. The reaction shown below is a(n):

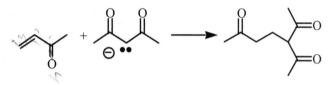

A.   1,2-conjugate addition
B.   1,4-conjugate addition
C.   electrophilic addition
D.   nucleophilic substitution

561. Aldolase is an enzyme that breaks down sugars. Fructose undergoes the reaction shown below. What type of reaction is catalyzed by aldolase?

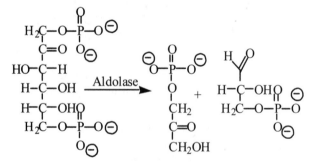

fructose 1,6-diphosphate

A.   dehydration
B.   hydration
C.   aldol condesation
D.   retro aldol condensation

Cis-jasmone is a perfume that can be synthesized from cis-8-undecene-2,5-dione as shown below. Use this synthesis to answer questions 562 - 566.

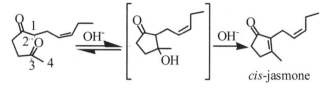

cis-jasmone

562. *Cis*-jasmone is a(n):

A.   enol
B.   α, β-unsaturated carbonyl
C.   aldol product
D.   enamine

563. The formation of the intermediate in the synthesis of *cis*-jasmone is a(n):

A.   base catalyzed aldol condensation.
B.   acid catalyzed aldol condensation.
C.   dehydration.
D.   conjugate addition.

564. The formation of *cis*-jasmone from the intermediate is a(n):

A.   base catalyzed aldol condensation.
B.   acid catalyzed aldol condensation.
C.   dehydration.
D.   conjugate addition.

565. Which hydrogen is removed in the first step of the *cis*-jasmone synthesis to create the enolate ion?

A.   The carbon labeled 1
B.   The carbon labeled 2
C.   The carbon labeled 3
D.   The carbon labeled 4

566. Which carbon is the electrophile in the first step of the *cis*-jasmone synthesis?

A.   The carbon labeled 1
B.   The carbon labeled 2
C.   The carbon labeled 3
D.   The carbon labeled 4

The iodoform test shown below is used to identify methyl ketones. A positive test gives the yellow crystalline CHI$_3$. Use the equation shown below to answer questions 567 - 569.

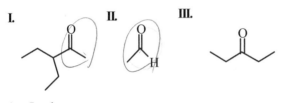

$$R-\overset{O}{\underset{}{C}}-CH_3 + 3I_2 + 4OH^- \longrightarrow R-\overset{O}{\underset{}{C}}-O^\ominus + CHI_3 + 3H_2O$$

567. Why does the hydroxide substitute for the carbon group only after three iodine ions have added?

   A. The presence of the iodide atoms forms a much better leaving group.
   B. Iodide is a good leaving group.
   C. HI is a strong acid.
   D. Water is a good leaving group.

568. Which of the following would be a good solvent for the iodoform test?

   A. acetone
   B. hexane
   C. water
   D. dichloromethane

569. Which of the following would produce a yellow solid in the iodoform test?

   I.   **II.**   **III.**

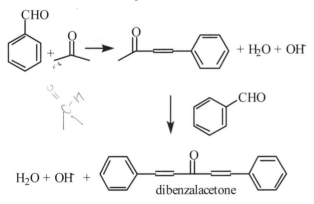

   A. I only
   B. II only
   C. I and II only
   D. I, II, and III

Dibezalacetone can be used as sunscreen because it has a large absorption band in the UV region. Use the synthesis shown below to answer questions 570 - 573.

570. The formation of benzalacetone is a(n):

   A. base catalyzed aldol condensation followed by dehydration.
   B. acid catalyzed aldol condensation followed by dehydration.
   C. dehydration followed by base catalyzed aldol condensation.
   D. dehydration followed by acid catalyzed aldol condensation.

571. In the formation of benzalacetone, acetone is the:

   A. nucleophile
   B. electrophile
   C. acid
   D. oxidant

572. In the formation of dibenzalacetone from benzalacetone, benzaldehyde is the:

   A. nucleophile
   B. electrophile
   C. base
   D. oxidant

573. Why does only one product form in the formation of benzalacetone?

   I.   the symmetry of acetone
   II. the absence of an alpha proton on benzaldehyde
   III. the absence of an alpha proton on acetone

   A. I only
   B. II only
   C. I and II only
   D. I, II, and III

574. Tautomerization can racemize optically active ketones and aldehydes. Which of the following would no longer rotate plane polarized light after reaction with acid?

   A. (S)-3-methyl-2-heptanone
   B. (R)-4-methyl-2-heptanone
   C. (S)-3-t-butylpentanal
   D. (R)-3-methylcyclopentanone

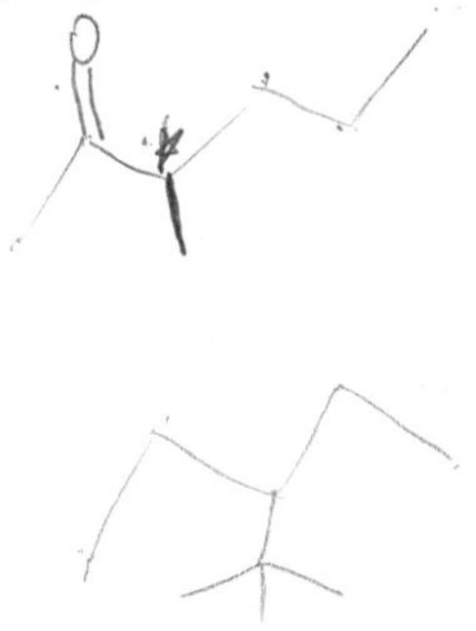

Use the scheme below to answer questions 575 - 582.

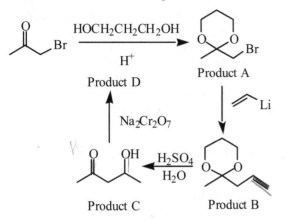

Product D

Na₂Cr₂O₇

Product A

Product C   Product B

H₂SO₄
H₂O

575. Product A is a(n):

A. hemiketal
B. hemiacetal
C. ketal
D. acetal

576. The reaction that forms product A from the starting compound is a(n):

A. nucleophilic substitution
B. electrophilic addition
C. conjugate addition
D. elimination

577. What is true about the two oxygen atoms found in product A?

A. One oxygen atom is from the diol and one is from the original carbonyl.
B. One oxygen atom is from water and one is from the diol.
C. Both oxygen atoms are from the diol.
D. Both oxygen atoms come from water.

578. How many carbonyl groups does product D have?

A. 0
B. 1
C. 2
D. 3

579. If product D is reacted with an excess of 2,2-dimethylbutanal in the presence of a base catalyst, what is true about the product?

A. A single aldol product is formed.
B. Two aldol products are formed.
C. Several aldol products are formed.
D. No product is formed.

580. If product D is reacted with LiAlH₄, then reacted with the 1-bromo-3-butanone and an acid catalyst, what is the product?

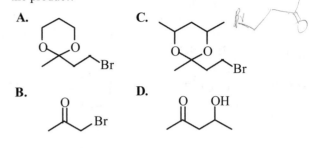

581. If the first step of the scheme were skipped, what would be the result?

A. The vinyl lithium will not react with the 1-bromo-3-propanone.
B. The vinyl lithium would also add to the carbonyl.
C. The same product would be obtained.
D. The reaction would proceed at a slower rate.

582. What would be the product if A was reacted with the reagent shown below, then with H₂SO₄ and H₂O?

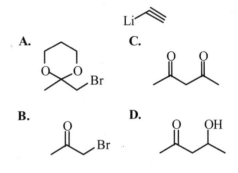

Use the scheme below to answer questions 583-592.

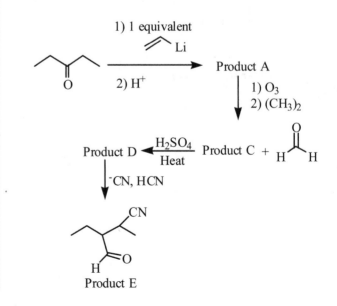

1) 1 equivalent

Li

2) H⁺

Product A

1) O₃
2) (CH₃)₂

Product D ← H₂SO₄ / Heat ← Product C + (H-CHO)

⁻CN, HCN

Product E

583. How many carbonyls are present in product A?

A. 0
B. 1
C. 2
D. 3

584. How many carbonyls are present in product C?

A. 0
B. 1
C. 2
D. 3

585. An excess of vinyl lithium was added to the 3-pentanone producing product B. If product B was then reacted under the same conditions as product A, how many carbonyls are present in the final product?

A. 0
B. 1
C. 2
D. 3

586. Where does the oxygen atom in product E come from?

A. the oxygen atom in 3-pentanone
B. an oxygen atom from O_3
C. an oxygen atom from H_2O
D. an oxygen atom from OH^-

587. Product D is a(n):

A. aldol product
B. α, β-unsaturated carbonyl
C. hydration product
D. cyano product

588. The transformation from product C to product D is a(n):

A. hydration.
B. dehydration.
C. aldol condensation.
D. hydrogenation.

589. If product D is reacted with $Na_2Cr_2O_7$, what is the product?

A. a primary alcohol
B. a secondary alcohol
C. a carboxylic acid
D. an aldehyde

590. If product D is hydrogenated with an excess of hydrogen, what is the product?

A. 2-ethyl-2-ene-butanal
B. 1-hydroxy-2-ethyl-2-butene
C. 2-ethylbutanal
D. 2-ethyl-1-butanol

591. The formation of product E from product D is a(n):

A. 1,2-conjugate addition
B. 1,4-conjugate addition
C. electrophilic addition
D. nucleophilic substitution

592. The formaldehyde byproduct from the second reaction in the scheme does not form an aldol product when reacted with a base catalyst. Why is no product formed?

A. Aldol condensations are not base catalyzed.
B. The alpha proton is not acidic.
C. Aldol condensations do not occur with aldehydes.
D. No enolate anion can form.

Diels-Alder reactions occur between a conjugated diene and a dienophile. Use the example of a Diels-Alder reaction shown below to answer questions 593 - 596.

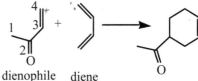

dienophile   diene

593. Which of the following is not a resonance structure of the diene?

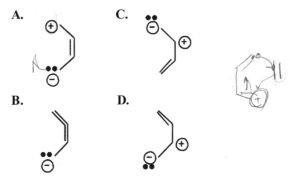

A.

C.

B.

D.

594. In the Diels-Alder reaction the nucleohpile is:

A. the diene
B. the dienophile
C. water
D. methanol

595. In order for a Diels-Alder to proceed at room temperature, the dienophile must be activated. Which carbon in the dienophile will be the site of the nulceophilic attack?

A. the carbon labeled 1
B. the carbon labeled 2
C. the carbon labeled 3
D. the carbon labeled 4

596. With respect to the diene, the reaction is a(n):

A. 1,2 conjugate addition
B. 1,4 conjugate addition
C. elimination
D. nucleophilic substitution

597. Which of the following aldol products would not form an α, β-unsaturated carbonyl?

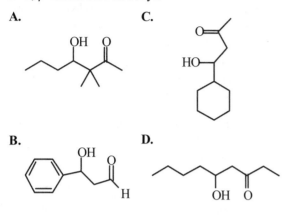

A.

C.

B.

D.

Carboxylic Acids and Derivatives

598. Which of the following has the highest boiling point?

A. formic acid
B. acetic acid
C. hexanoic acid
D. octanoic acid

599. Which of the following has the highest boiling point?

A. propane
B. proponic acid
C. 1-propanol
D. propanal

Use the table below to answer questions 600-601.

Acid	Boiling point
butanoic acid	164°C
heptanoic acid	223°C
octanoic acid	239°C
decanoic acid	270°C

600. What is the best explanation for the boiling point trend observed in the table?

A. Increased molecular weight increases hydrogen bonding.
B. Increased molecular weight increases London Dispersion forces.
C. Increased molecular weight decreases hydrogen bonding.
D. Increased molecular weight decreases London Dispersion forces.

601. A scientist finds a bottle labeled caproic acid. She determines that it is a carboxylic acid and that the boiling point of the liquid is found to be 205 °C. Which of the following is the best assignment of the acid?

A. butanoic acid
B. hexanoic acid
C. nonanoic acid
D. dodecanoic acid

602. Which of the following has the greatest solubility in water?

A. pentanoic acid
B. pentane
C. 1-pentanol
D. pentanal

603. Why is propanoic acid miscible in water when octanoic acid is only 0.7% soluble?

A. Propanoic acid has a longer alkyl chain than octanoic acid increasing solubility.
B. Propanoic acid has a longer alkyl chain than octanoic acid decreasing solubility.
C. Propanoic acid has a shorter alkyl chain than octanoic acid increasing solubility.
D. Propanoic acid has a shorter alkyl chain than octanoic acid decreasing solubility.

604. Which of the following has the most acidic proton?

A. pentanoic acid
B. pentane
C. 1-pentanol
D. pentanal

605. Which of the following has the most acidic proton?

A. pentanoic acid
B. 2-hydroxypentanoic acid
C. 2-methylpentanoic acid
D. 2-nitropentanoic acid

606. Which of the following has the most acidic proton?

A. 2-butanone
B. butanal
C. butanoyl chloride
D. methyl butanoate

607. Why does p-nitrobenzoic acid have a pK_a of 3.41 when p-chlorobenzoic acid has a pK_a value of 3.98?

   A. p-nitrobenzoic acid has an electron donating group that further stabilizes the anion.
   B. p-chlorobenzoic acid has an electron donating group that further stabilizes the anion.
   C. p-chlorobenzoic acid has an electron withdrawing group that further stabilizes the anion.
   D. p-nitrobenzoic acid has an electron withdrawing group that further stabilizes the anion.

608. Why does ethanoic acid have a pK_a of 4.74 when ethanol has a pK_a value of 15.9?

   A. Ethanol hydrogen-bonds with water.
   B. Ethanoic acid has a second oxygen atom that contributes to resonance stabilization of the anion.
   C. Ethanol has a second oxygen atom that contributes to resonance stabilization of the anion.
   D. Ethanoic acid has stronger hydrogen bonding.

Use the table shown below to answer questions 609 - 611.

Acid	pK_a
CH_3CH_2COOH	4.87
CH_3COOH	4.74
$CH_3CH_2CH_2COOH$	4.82
$ClCH_2CH_2CH_2COOH$	4.52
$ClCH_2COOH$	2.86
Cl_3CCOOH	0.64

609. Which of the following has the most acidic proton?

   A. CH_3CH_2COOH
   B. CH_3COOH
   C. $ClCH_2COOH$
   D. Cl_3CCOOH

610. Why does $ClCH_2CH_2CH_2COOH$ have a greater pK_a than $ClCH_2COOH$?

   A. 4-chlorobutanoic acid has a higher molecular weight than 2-chloroethanoic acid
   B. 4-chlorobutanoic acid has a lower molecular weight than 2-chloroethanoic acid
   C. 4-chlorobutanoic acid has an electron withdrawing group on the gamma carbon while 2-chloroethanoic acid has an electron withdrawning group on the alpha carbon
   D. 4-chlorobutanoic acid is a longer chain than 2-chloroethanoic acid

611. What would be the expected pK_a of 2,2-dichloroethanoic acid?

   A. 4.87
   B. 2.86
   C. 1.26
   D. 0.64

612. Both carbon-oxygen bonds in formate are 1.26Å compared to 1.20 Å for a carbon-oxygen double bond and 1.34 Å for a carbon-oxygen single bond. What is the best explanation for this observation?

   A. The negative charge shortens the carbon oxygen single bond.
   B. The negative charge lengthens the carbon oxygen double bond.
   C. The negative charge creates a larger dipole moment.
   D. The actual structure is a combination of two significant resonance structures.

613. What is the result of the reaction of an alcohol and an acid chloride?

   A. ester
   B. ether
   C. amide
   D. imide

614. What is the result of the reaction of an amine and a carboxylic acid?

   A. ester
   B. ether
   C. amide
   D. imide

615. If ethyl alcohol ferments in the presence of oxygen (an oxidant) vinegar is formed. What is vinegar?

   A. ethanoic acid
   B. diethyl ether
   C. ethanol
   D. ethane

616. Which of the following reagents will convert butanoic acid into butyl chloride?

   A. Cl_2
   B. HCl
   C. $SOCl_2$
   D. $LiAlH_4$

617. Which of the following will form acetic anhydride when reacted with the acetate ion?

   A. acetyl chloride
   B. acetic acid
   C. methyl acetate
   D. N-ethylacetamide

618. Which of the following will react with acetic acid to form ethyl acetate?

   A. acetic anhydride
   B. acetic acid
   C. ethanol
   D. ethyl chloride

619. What is the source of the circled oxygen atom below in the ethyl acetate produced from the reaction described in question 618?

   A. acetic acid
   B. ethanol
   C. water
   D. oxygen

620. A scientist was synthesizing 3-oxo-pentaonic acid when bubbling was observed at room temperature. What is the cause of the bubbling?

   A. release of O_2
   B. release of H_2
   C. release of CO_2
   D. release of H_2O vapor

Luminol is used in the detection of blood. Use the scheme for the synthesis of luminol (shown below) to answer questions 621 - 624.

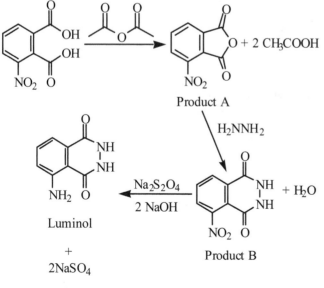

621. Product A is a(n):

   A. ester
   B. amide
   C. anhydride
   D. ether

622. What is the reason for using product A as an intermediate in the formation of product B?

   A. Carboxylic acids are more reactive than anhydrides.
   B. Anhydrides are more reactive than carboxylic acids.
   C. Five membered rings are more reactive.
   D. Five membered rings are less reactive.

623. The transformation of product A to product B is a(n):

   A. oxidation
   B. reduction
   C. nucleophilic substitution
   D. nucleophilic addition

624. The transformation of product B to lumniol is a(n):

   A. oxidation
   B. reduction
   C. nucleophilic substitution
   D. nucleophilic addition

625. The formation of an ester by the reaction of a carboxylic acid and an alcohol is reversible. How can the yield of the ester be increased?

   I.   Increase the concentration of alcohol.
   II. Substitute the appropriate anhydride for the carboxylic acid.
   III. Add an acid catalyst.

   A. I only
   B. III only
   C. I and III only
   D. I, II, and III

626. The synthesis of methyl benzoate from benzoic acid and methanol requires an acid catalyst. However, methyl benzoate can be synthesized from benzoyl chloride and methanol without a catalyst. Why are different conditions required for these two reactions?

   A. The product is more stable in the acid chloride reaction.
   B. The product is more stable in the carboxylic acid reaction.
   C. Acid chlorides are less reactive than carboxylic acids.
   D. Acid chlorides are more reactive than carboxylic acids.

627. What does the acid catalyst do in the synthesis of an ester from a carboxylic acid and alcohol?

    I. creates a better leaving group
    II. activates the nucleophile
    III. activates the electrophile

    A. I only
    B. III only
    C. I and III only
    D. I, II, and III

The synthesis of triphenylcarbinol using a Grignard reagent is shown below. Use the scheme to answer questions 628 - 631.

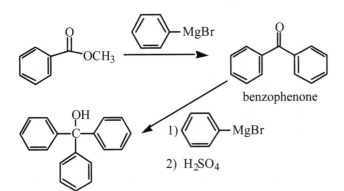

triphenylcarbinol

628. The formation of benzophenone is a(n):

    A. elimination
    B. conjugate addition
    C. nucleophilic substitution
    D. nucleophilic addition

629. If only one equivalent of phenyl magnesium bromide is added to the reaction, 0.5 equivalents of triphenylcarbinol forms and no benzophenone. Given these results what can be concluded about benzophenone?

    A. Benzophenone is more reactive than methyl benzoate.
    B. Methyl benzoate is more reactive than benzophenone.
    C. Benzophenone has a more conjugated system than methyl benzoate.
    D. Benzophenone has a less conjugated system than methyl benzoate.

630. The reaction of $Na_2Cr_2O_7$ in sodium hydroxide with triphenylcarbinol will produce a(n):

    A. anion
    B. ketone
    C. aldehyde
    D. carboxylic acid

631. Which of the following would be the most acidic?

    A. ethanol
    B. benzophenone
    C. methyl benzoate
    D. triphenylcarbinol

Use the synthesis below to answer questions 632 - 635.

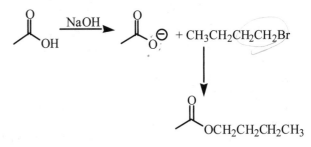

632. In the synthesis of butyl acetate, the acetate ion is a(n):

    A. nucleophile
    B. electrophile
    C. catalyst
    D. solvent

633. If (S)-2-bromobutane is used in place of the 1-bromobutane, what is true about the ester that is formed?

    A. The ester will be optically active.
    B. The ester will be optically inactive.
    C. No product will form.
    D. It will form faster than the butyl acetate.

634. What other reaction will compete with the ester product?

    A. dehydration
    B. dehydrohalogenation
    C. hydration
    D. decarboxylation

635. Which of the following would increase the rate of the reaction?

    I. Increase the concentration of acetate ion.
    II. Increase the concentration of the alkyl halide.
    III. Increase the temperature.

    A. I only
    B. I and III only
    C. II and III only
    D. I, II, and III

636. The structure of N,N-diethyl-meta-toluamide (DEET) is shown below. DEET is one of the best insect repellents. What two reagents would be best to use in order to synthesize DEET?

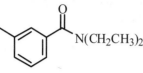

A. meta-methylbenzoyl chloride and ammonia
B. meta-methylbenzoyl chloride and diethyl amine
C. meta-methylbenzoic acid and diethyl amine
D. meta-methylbenzoic acid and ammonia

637. Which of the following will lose CO_2 at room temperature?

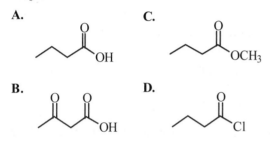

A reaction for forming derivatives of unknown acids for the purpose of identification is shown below. Use this scheme to answer questions 638 - 642.

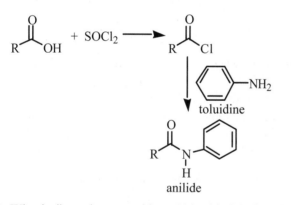

638. Why is the unknown acid reacted with $SOCl_2$ before reacting with toluidine?

A. Acid chlorides are more reactive than carboxylic acids.
B. Carboxylic acids are more reactive than acid chlorides.
C. The carbonyl of a carboxylic acid is more electrophilic than the carbonyl of an acid chloride.
D. An acid chloride catalyzes the formation of an anilide.

639. An anilide is an:

A. amine
B. amide
C. imine
D. ester

640. The formation of the acid chloride is a(n):

A. elimination
B. conjugate addition
C. nucleophilic substitution
D. nucleophilic addition

641. In the formation of the anilide, toluidine is a(n):

A. eletrophile
B. nucleophlie
C. acid
D. catalyst

642. The formation of the anilide is a(n):

A. elimination
B. conjugate addition
C. nucleophilic substitution
D. nucleophilic addition

643. Which of the following acids would be most reactive with the $SOCl_2$?

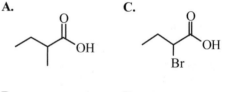

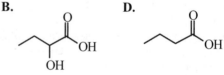

Bis(2,4,6-trichlorophenyl)oxalate is used in light sticks. When hydrogen peroxide and a flouorescer are added to this molecule, light is produced. Use the synthesis reaction for bis(2,4,6-trichlorophenyl)oxalate to answer questions 644 - 649.

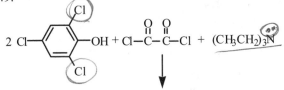

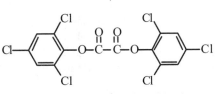

Bis(2,4,6-trichlorophenyl)oxalate

644. Why is triethylamine added to the reaction?

A. to form an amide product
B. to act as a base
C. to lower the pH of the reaction mixture
D. to be the solvent

645. The proton on the phenol must be removed to start the reaction. Why is the phenol proton acidic?

I. Electron withdrawing groups are present.
II. The anion is stabilized by resonance.
III. The phenol is a carboxylic acid.

A. I only
B. II only
C. I and II only
D. I and III only

646. Bis(2,4,6-trichlorophenyl)oxalate is a(n):

A. amide
B. ester
C. acid chloride
D. carboxylic acid

647. In the synthesis of bis(2,4,6-trichlorophenyl)oxalate, 2,4,6-trichlorophenol is a(n):

A. nucleophile
B. electrophile
C. base
D. reductant

648. What does the prefix "bis" in bis (2,4,6 trichlorophenyl) oxalate indicate?

A. more than one chloride ion
B. two 2,4,6-trichlorophenyloxalate groups
C. two moles of product
D. two oxygens

649. In the reaction that produces the energy needed for light emission (shown below), the intermediate is:

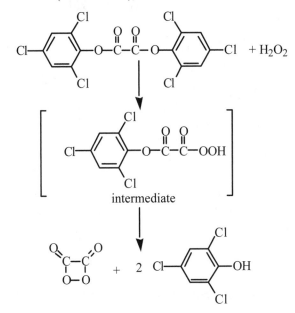

intermediate

A. a stable molecule that is easy to isolate
B. at a higher oxidation state than bis (2,4,6-trichlorophenyl) oxalate
C. at a lower oxidation state than bis (2,4,6-trichlorophenyl) oxalate
D. the same as a transition state

650. E-α-phenylcinnamic acid (shown below) undergoes decarboxylation to form Z-stilbene when reacted with a copper chromite catalyst and quinoline. What is the structure of Z-stilbene?

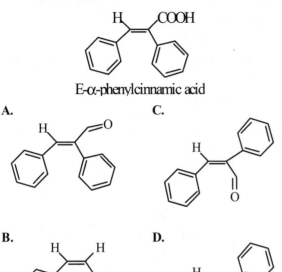

E-α-phenylcinnamic acid

A.

C.

B.

D.

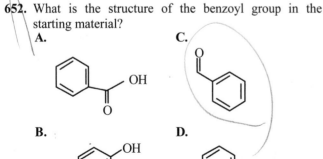

Anthraquinone is synthesized from 2-benzoylbenzoic acid. Refer to the synthesis below to answer questions 651 - 654.

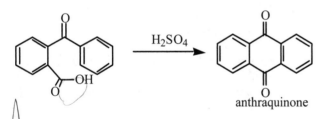

anthraquinone

651. The synthesis of anthraquinone from 2-benzoylbenzoic acid is a(n):

A. oxidation
B. reduction
C. acid/base reaction
D. decarboxylation

652. What is the structure of the benzoyl group in the starting material?

A.

C.

B.

D.

653. In the synthesis of anthraquinone, the carboxylic acid group is:

A. the nucleophile
B. the electrophile
C. the catalyst
D. the oxidant

654. In the synthesis of anthraquinone, the H_2SO_4 is:

A. the nucleophile
B. the electrophile
C. the catalyst
D. the oxidant

In the reaction of an alcohol and a carboxylic acid, a tetrahedral intermediate is formed. The intermediate can either return to the carboxylic acid or convert to an ester. Use this information to answer questions 655 - 658.

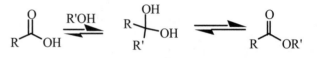

655. What is the product when acetic acid is reacted with 1-butanol in the presence of an acid catalyst?

A.

C.

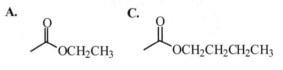

B.

D.

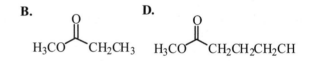

656. If a scientist wished to obtain a methyl ester from a carboxylic acid, what reaction conditions would favor the product?

A. adding the carboxylic acid to acidic methanol
B. adding methanol and the carboxylic acid to hot water
C. adding methanol and the carboxylic acid to hot acidic water
D. adding methanol and the carboxylic acid to cold acidic water

657. When the tetrahedral intermediate forms the ester, what is the leaving group?

A. methoxide
B. methanol
C. hydroxide
D. water

658. When the tetrahedral intermediate returns to the carboxylic acid, what is the leaving group?

A. an alkoxide
B. an alcohol
C. hydroxide
D. water

Adipic acid is a monomer used in the formation of Nylon 6.6. Use the synthesis of adipic acid shown below to answer questions 659 - 667.

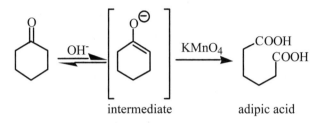

intermediate   adipic acid

659. The formation of the intermediate from cyclohexanone is:

A. a nuclephilic substitution.
B. an electrophilic addition
C. a tautomerzation.
D. resonance stabilization.

660. The formation of adipic acid from cyclohexanone is a(n):

A. reduction
B. oxidation
C. elimination
D. substitution

661. Adipic acid is a(n):

A. ketone
B. ester
C. di ester
D. di carboxylic acid

662. Which of the following reagents will convert adipic acid into a diol?

A. $LiAlH_4$
B. PBr_3
C. $Na_2Cr_2O_7$
D. O_3

663. Which of the following reagents will convert adipic acid into a diester?

A. water
B. methanol
C. acetic acid
D. $Na_2Cr_2O_7$

664. What would be the result if no base and only $KMnO_4$ was added to the cyclohexanone?

A. formation of a mono carboxylic acid
B. formation of a di carboxylic acid
C. formation of an alcohol
D. no reaction

665. The polymerization reaction with adipic acid to form Nylon 6.6 is shown below. Nylon 6.6 is a(n)

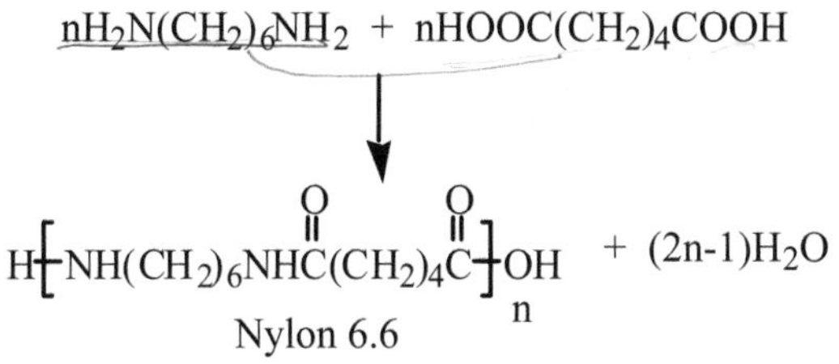

$$nH_2N(CH_2)_6NH_2 \ + \ nHOOC(CH_2)_4COOH$$

$$H\text{--}[NH(CH_2)_6NHC(CH_2)_4C]\text{--}OH \ + \ (2n\text{-}1)H_2O$$
$$ \overset{O}{\overset{||}{}} \overset{O}{\overset{||}{}} _n$$
Nylon 6.6

A. polyester
B. polyamide
C. dicarboxylic acid
D. diamine

666. In the polymerization reaction shown in question 665, what are the hexmethylenediamine and adipic acid respectively?

A. nucleophile, electrophile
B. electrophile, nucelophile
C. acid, base
D. solvent, catalyst

667. Polyethylene terephthalate (Dacron) is shown below. How does the stability of Nylon 6.6 (question 665) against hydrolysis compare to Dacron's stability against hydrolysis?

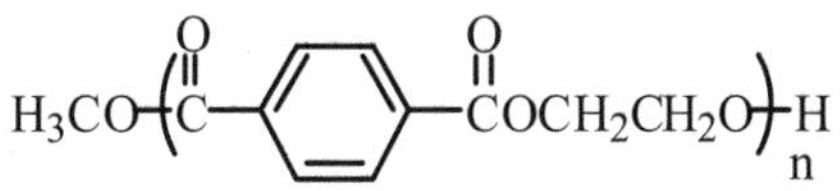

$$H_3CO\text{--}[C\text{---}\bigcirc\text{---}COCH_2CH_2O]\text{--}H$$
$$\overset{O}{\overset{||}{}} \overset{O}{\overset{||}{}} _n$$

A. Nylon 6.6 is more reactive than Dacron.
B. Nylon 6.6 is less reactive than Dacron.
C. Nylon 6.6 and Dacron have the same reactivity.
D. Relative stability cannot be determined.

S(+)-Ethyl 3-hydroxybutanoate is the product of ethyl acetoacetate's reaction with yeast. This product can be further reacted with 3,5-dinitrobenzoyl chloride in the presence of pyridine to give 3,5-dinitrobenzoate (shown below). Use this synthesis to answer questions 667 – 672.

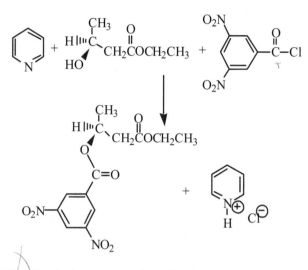

668. What is the purpose of pyridine in the synthesis?

    A. an oxidant
    B. a reductant
    C. acid catalyst
    D. base catyalst

669. Why would this synthesis be complicated if 3,5-dinitrobenzoic acid were used instead of 3,5-dinitrobenzoyl chloride?

    A. Carboxylic acids cannot undergo nucleophilic substitution.
    B. The proton of the hydroxyl group of the carboxylic acid would be more acidic than the proton on the alcohol.
    C. Acid chlorides are less reactive than carboxylic acids.
    D. The carboxylic acid would react with the ester functionality in the ethyl 3-hydroxybutanoate.

670. How does the reactivity of the carbonyl in 3,5-dintrobenzoyl chloride compare to the carbonyl in ethyl 3-hydroxybutanoate?

    A. The carbonyl in 3,5-dinitrobenzoyl is more reactive than the carbonyl in ethyl 3-hydroxybutanoate.
    B. The carbonyl in 3,5-dinitrobenzoyl is less reactive than the carbonyl in ethyl 3-hydroxybutanoate.
    C. They have the same reactivity.
    D. Relative stability cannot be determined.

671. How do the nitro groups on the benzene ring in 3,5-dinitrobenzoyl chloride affect the reaction?

    A. They inhibit the reaction because they are deactivating.
    B. They inhibit the reaction because they are electron withdrawing.
    C. The enhance the reaction because they activate the nucleophile.
    D. They enhance the reaction because they activate the electrophile.

672. Why does the stereochemistry of the molecule not change with the nucleophilic substitution?

    A. The reaction goes by a S_N2 mechanism.
    B. The reaction goes by a S_N1 mechanism.
    C. The substitution does not take place at the chiral carbon.
    D. Nucleophilic substitution of carboxylic acid derivatives goes with the retention of stereochemistry.

Phenolphthalein is an indicator for pH. In solutions of pH greater than 8, phenolphthalein is pink. Use the reaction of phenolphthalein shown below to answer question 673 - 676.

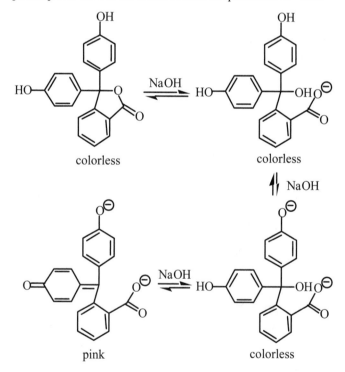

673. At low pH phenolphthalein is a(n):

    A. ester
    B. ether
    C. carboxylic acid
    D. anhydride

674. The reaction with the first addition of an equivalent of NaOH to phenolphthalein is a(n):

A. hydrolysis
B. dehydration
C. hydrogenation
D. decarboxylation

675. The last reaction when the final equivalent of NaOH is added is a(n):

A. hydrolysis
B. dehydration
C. hydrogenation
D. decarboxylation

676. In a solution with a pH value greater than 8 phenolphthalein is a(n):

A. ester
B. ether
C. carboxylic acid
D. anhydride

Use the reaction scheme below to answer questions 677 - 684.

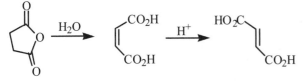

maleic anhydride   maleic acid   fumaric acid

677. The formation of maleic acid from maleic anhydride is a(n):

A. isomerization
B. hydrolysis
C. carboxylation
D. decarboxylation

678. The formation of fumaric acid from maleic acid is a(n):

A. isomerization
B. hydrolysis
C. carboxylation
D. decarboxylation

679. In the formation of maleic acid from maleic anhydride, water is a(n):

A. catalyst
B. base
C. nucleophile
D. electrophile

680. Maleic acid and fumaric acid are:

A. diastereomers
B. enantiomers
C. structural isomers
D. meso

681. How does the stability of maleic acid compare to fumaric acid?

A. Maleic acid is more stable than fumaric acid.
B. Fumaric is more stable than maleic acid.
C. The stabilites are the same.
D. The relative stabilities cannot be determined.

682. Which of the following would result from the reaction of fumaric acid with an excess of LiAlH$_4$?

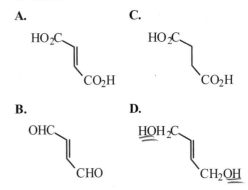

683. Which of the following reagents would produce two equivalents of an alpha carbonyl carboxylic acid from maleic acid?

A. H$^+$, water
B. NaBH$_4$
C. KMnO$_4$
D. O$_3$, (CH$_3$)$_2$S

684. Which of the following would be more reactive with diethylamine to form an amide?

A. maleic anhydride
B. maleic acid
C. fumaric acid
D. water

Acetic acid can by synthesized by the following method. Use this synthesis to answer questions 685 - 686.

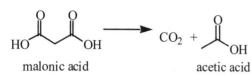

malonic acid   acetic acid

685. Why is the formation of acetic acid at 150°C not reversible?

  A. The reaction is endothermic as written.
  B. The CO_2 bubbles out of the solution.
  C. The reactant is more stable than the product.
  D. There is a large activation energy for the forward reaction.

686. Why does adipic acid (question 659) require harsher conditions to lose carbon dioxide than malonic acid?

  A. Malonic acid is a dicarboxylic acid.
  B. Malonic acid has a larger pK_a than adipic acid.
  C. Malonic acid has a smaller pK_a than adipic acid.
  D. Malonic acid has a carbonyl on the beta carbon.

Isoamyl propionate is one of the primary components in the pineapple odor. Use the reaction shown below to answer questions 687 - 690.

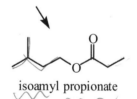

propionic anhydride   isoamyl alcohol

isoamyl propionate

687. The synthesis is catalyzed by an acid. What does the acid do in the first step of the synthesis?

  A. activate the electrophile by protonating the propionic anhydride
  B. activate the nucleophile by protonating the propionic anhydride
  C. activate the electrophile by protonating the isoamyl alcohol
  D. activate the nucleophile by protonating the isoamyl alcohol

688. Which of the following reagents would be more reactive with isoamyl alcohol?

  A. propionic acid
  B. propionyl chloride
  C. methyl propionate
  D. propyl acetate

689. The anhydride is present in excess during the reaction. As part of the workup, the reaction mixture is washed with a weak base. What does the weak base do?

  A. hydrolyzes the isoamyl propionate
  B. hydrolyzes the propionic anhydride
  C. deprotonates the isoamyl alcohol
  D. deprotonates the propionic anhydride

690. As described in question 689, a weak base is used during the workup of the isoamyl propionate. Why is a strong base avoided?

  A. to avoid hydrolyzing the isoamyl propionate
  B. to avoid hydrolyzing the propionic anhydride
  C. to avoid deprotonating the isoamyl alcohol
  D. to avoid deprotonating the propionic anhydride

Amines

691. The structure of piperazine, used to kill intestinal worms, is shown below. Piperazine is a:

  A. primary amine
  B. secondary amine
  C. tertiary amine
  D. quaternary amine

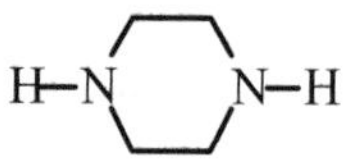

692. The structure of amphetamine, used as a stimulant, is shown below. Amphetamine is a:

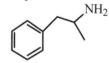

  A. primary amine
  B. secondary amine
  C. tertiary amine
  D. quaternary amine

693. The neurotransmitter shown below contains what functional groups?

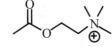

A. a primary amine and a carboxylic acid
B. a tertiary amine and a carboxylic acid
C. a tertiary amine and an ester
D. a quaternary amine and an ester

694. N,N-diethylaniline is a:

A. primary amine
B. secondary amine
C. tertiary amine
D. quaternary amine

695. What is the correct structure of dimethylethylamine?

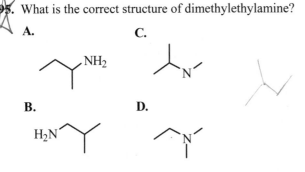

696. Putrescine (1,4-aminobutane) has an odor that matches the name. What is the structure of putrescine?

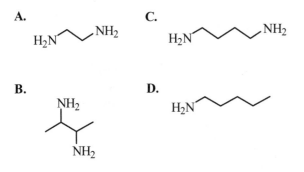

697. The amine shown below would rotate plane polarized light:

A. clockwise
B. counterclockwise
C. The direction cannot be determined.
D. The molecule will not rotate plane polarized light.

698. The amine shown on the left is optically active, while the amine shown on the right is not. What is the best explanation for this observation?

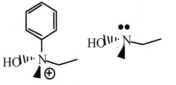

A. The amine on the left has a phenyl group and the amine on the right does not.
B. The nitrogen on the left has resonance structures.
C. The amine on the right has a lone pair of electrons so nitrogen inversion is not possible.
D. The amine on the left does not have a lone pair of electrons so nitrogen inversion is not possible.

699. Which of the following is a true statement about the two amines shown below?

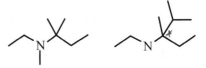

A. The amine on the left is optically active and the amine on the right is not optically active.
B. The amine on the right is optically active and the amine on the left is not optically active.
C. Both are optically active.
D. Both are not optically active.

700. Which vector best represents the net dipole moment in the amine shown below?

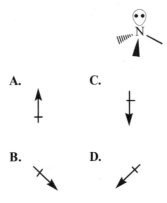

A.

C.

B.

D.

701. Which of the following will have hydrogen bonding?

    I.   triethylamine
   II. diethylamine
  III. ethylamine

    A. I only
    B. III only
    C. II and III only
    D. I, II, and III

702. Which of the following will have hydrogen bonding with water in an aqueous solution?

    I.   triethylamine
   II. diethylamine
  III. ethylamine

    A. I only
    B. III only
    C. II and III only
    D. I, II, and III

703. Which of the following would be the most soluble in water?

    A. ethylamine
    B. trimethylamine
    C. triphenylamine
    D. N-phenyldiethylamine

704. Which of the following would be the most soluble in water?

    A. cyclohexylamine
    B. cyclohexane
    C. benzene
    D. methyl phenyl ether

705. Why is the boiling point of trimethylamine 3°C, while n-propylamine is 48°C?

    A. Trimethylamine has a higher molecular weight than n-propylamine.
    B. Trimethylamine has a lower molecular weight than n-propylamine.
    C. Trimethylamine does hydrogen bond and n-propylamine does not hydrogen bond.
    D. Trimethylamine does not hydrogen bond and n-propylamine does hydrogen bond.

706. Why does ethylamine have a boiling point of 17°C, while ethanol has a boiling point of 78°C?

    A. Ethylamine has a higher molecular weight than ethanol.
    B. Ethylamine has a lower molecular weight than ethanol.
    C. Oxygen is more electronegative than nitrogen leading to stronger hydrogen bonding in ethanol.
    D. Oxygen is more electronegative than nitrogen leading to stronger hydrogen bonding in ethylamine.

707. Which of the following has the lowest boiling point?

    A. n-propylamine
    B. isopropylamine
    C. ethylmethylamine
    D. triethylamine

708. When diethylamine is dissolved in water, what is true about the pH of the solution?

    A. $pH > 7$
    B. $pH = 7$
    C. $pH < 7$
    D. The relative pH value cannot be predicted.

709. In the following reaction ammonia is a(n):

$$Fe^{3+} + 6\,NH_3 \longrightarrow \begin{bmatrix} H_3N\text{-}Fe\text{-}NH_3 \\ H_3N \quad NH_3 \\ NH_3 \quad NH_3 \end{bmatrix}^{3+}$$

    A. Lewis acid
    B. Lewis base
    C. reductant
    D. oxidant

Use the table below to answer questions 710-715.

Amine	pK_b
aniline	9.40
N,N-dimethylaniline	8.94
p-toluidine	8.92
p-fluoroaniline	9.36
p-chloroaniline	10.00
p-iodoaniline	10.22
p-nitroaniline	13.00

710. Given the structures for p-nitroaniline and p-toluidine below, what is the best explanation for the relative pK_b values for the amines in the table?

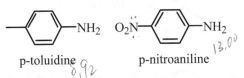

p-toluidine   p-nitroaniline

A. *Para*-toluidine is more basic than *p*-nitroaniline because a methyl group is electron donating and a nitro group is electron withdrawing.
B. *Para*-toluidine is less basic than *p*-nitroaniline because a methyl group is electron donating and a nitro group is electron withdrawing.
C. *Para*-toluidine is more basic than *p*-nitroaniline because a methyl group is electron withdrawing and a nitro group is electron donating.
D. *Para*-toluidine is less basic than *p*-nitroaniline because a methyl group is electron withdrawing and a nitro group is electron donating.

711. Given the structures in question 710, what is the structure of N-methylaniline?

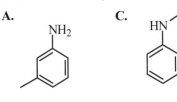

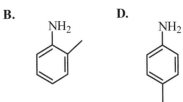

712. Why is the pK_b of aniline higher than the pK_b of N,N-methylaniline?

A. N,N-methylaniline has a higher molecular weight than aniline.
B. N,N-methylaniline has a lower molecular weight than aniline.
C. Tertiary amines are more basic than primary amines when the substituents are electron donating.
D. Tertiary amine are less basic than primary amines when the substituents are electron donating.

713. What is the approximate pK_b value for *p*-bromoaniline?

A. 9.36
B. 9.56
C. 10.15
D. 10.32

714. What is the approximate pK_b value for *p*-anisidine (4-methoxyaniline)?

A. 8.70
B. 9.60
C. 10.00
D. 13.20

715. What is the approximate pK_b value for N-methylaniline?

A. 8.75
B. 9.21
C. 9.70
D. 10.35

716. The reaction shown below favors the reactants. What is the best explanation for the energy relationship between the reactants and the products?

pyrrole

A. Secondary amines are more basic than primary amines.
B. Primary amines are more basic than secondary amines.
C. The product is stabilized by aromaticity.
D. The reactant is stabilized by aromaticity.

717. In the reaction shown in question 716, what is true about pyrrole?

I. It is a Bronsted-Lowry base.
II. It is a Lewis base.
III. It is a Lewis acid.

A. III only
B. II only
C. I and II only
D. I, II, and III

Use the reaction shown below to answer questions 718–720.

718. The amine group on the phenyl ring is:
A. ortho, para directing and activating.
B. ortho, para directing and deactivating.
C. meta directing and activating.
D. meta directing and deactivating.

719. What is the purpose of the $NaHCO_3$?

   A. neutralize the HBr that is formed
   B. neutralize the amine group
   C. protonate the amine group
   D. reduce the amine group

720. Why are there three substitutions of bromine for hydrogen?

   A. The amine group activates three sites.
   B. The amine group deactivates three sites.
   C. There is not enough Br_2 to do five substitutions.
   D. There is not enough Br_2 to replace the amine group.

721. When ethyl amine is reacted with an excess of methyl iodide, how many equivalents of methyl iodide will react with one equivalent of ethyl amine?

   A. 1 equivalent
   B. 2 equivalents
   C. 3 equivalents
   D. 4 equivalents

722. When diethyl amine is reacted with an excess of methyl iodide, how many equivalents of methyl iodide will react with one equivalent diethyl amine?

   A. 1 equivalent
   B. 2 equivalents
   C. 3 equivalents
   D. 4 equivalents

723. If a methyl ethyl amine was desired from the reaction of ethyl amine with methyl iodide, what could be done to ensure single alkylation by methyl iodide?

   A. Add the methyl iodide slowly to a large excess of ethyl amine.
   B. Add the ethyl amine slowly to a large excess of methyl iodide.
   C. Add the ethyl amine quickly to a large excess of methyl iodide.
   D. Add one equivalent of both reactants.

724. In the reaction of ethylamine with methyl iodide, ethylamine is a(n):

   A. electrophile
   B. nucleophile
   C. acid
   D. catalyst

725. The product of the reaction of an excess of methyl iodide with ethylamine is a(n):

   A. primary amine
   B. secondary amine
   C. tertiary amine
   D. quaternary amine

726. Which of the following amines would react with acetone to form an imine?

   A. ethylamine
   B. dimethylamine
   C. triphenylamine
   D. N,N-diphenylethylamine

727. Which of the following amines would react with acetone to form an enamine?

   A. ethylamine
   B. dimethylamine
   C. triphenylamine
   D. N,N-diphenylethylamine

Use the reaction shown below to answer questions 728-731.

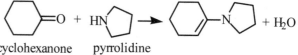

cyclohexanone   pyrrolidine

728. The product of the reaction of cyclohexanone with pyrrolidine is a(n):

   A. imine
   B. enamine
   C. aromatic amine
   D. quantanary amine

729. If butylamine was substituted for pyrrolidine, the product would be a(n):

   A. imine
   B. enamine
   C. aromatic amine
   D. quaternary amine

730. The reaction of cyclohexanone with pyrrolidine is a(n):

   A. nucleophilic addition.
   B. nucleophilic addition then dehydration.
   C. nucleophilic substitution.
   D. nucleophilic substitution then dehydration.

731. In the reaction of cyclohexanone with pyrrolidine, pyrrolidine is a(n):

   A. electrophile
   B. nucleophile
   C. acid
   D. catalyst

Use the scheme for the Hindsberg test of unknown amine complexes to answer questions 732-739.

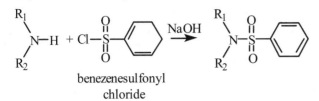

benezenesulfonyl
chloride

732. Benzenesulfonyl chloride is analogous to a(n):

   A. ester
   B. acid chloride
   C. imine
   D. amine

733. If R_1 is a hydrogen atom, then the products of the reaction with bezenesulfonyl chloride are typically soluble in water. However, when R_1 is an alkyl group the products are insoluble. What is the best explanation for this result?

   A. The primary amine forms a product with an acidic proton which forms a salt in the basic reaction mixture.
   B. The secondary amine forms a product with an acidic proton which forms a salt in the basic reaction mixture.
   C. Secondary amines are more reactive than primary amines.
   D. Tertiary amines are more reactive than secondary amines.

734. The product of the reaction of benzenesulfonyl chloride with the amine is a(n):

   A. benzenesulfonamide
   B. benzenesulfonate
   C. benzenesulfonyl enamine
   D. benzenesulfonyl ester

735. Why do tertiary amines not react with benzenesulfonyl chloride?

   A. Tertiary amines have molecular weights that are too high.
   B. Tertiary amines are not basic.
   C. Tertiary amines do not have a lone pair of electrons.
   D. Tertiary amines do not have an acidic hydrogen necessary to form the final product.

736. What is the product if dimethylamine is reacted with benzenesulfonyl chloride?

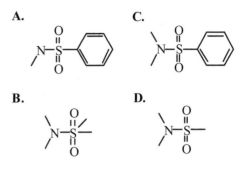

737. In the Hindsberg test, the benzenesulfonyl chloride is a(n):

   A. electrophile
   B. nucleophile
   C. base
   D. catalyst

738. If benzoyl chloride was substituted for benzenesulfonyl chloride in the Hindsberg reaction, what would the product be when reacted with butyl amine?

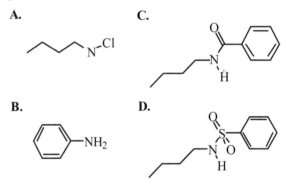

739. The product from the reaction of benzoyl chloride with butyl amine is not soluble in the alkali solution, but the product of butyl amine and benzenesulfonyl chloride is soluble. Why is the first product not soluble, but the second product is soluble?

   A. Benzoyl chloride produces a product that is less acidic because it has less resonance stabilization of a negative charge.
   B. Benzoyl chloride produces a product that is more acidic because it has less resonance stabilization of a negative charge.
   C. Benzoyl chloride produces a product that is less acidic because it has more resonance stabilization of a negative charge.
   D. Benzoyl chloride produces a product that is more acidic because it has more resonance stabilization of a negative charge.

Refer to the following synthesis to answer questions 740-744.

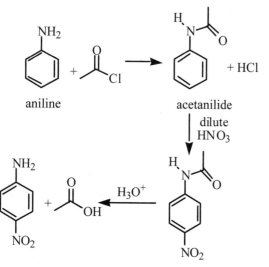

740. Acetanilide is a(n):

A. imine
B. enamine
C. amide
D. quaternary amine

741. Which of the following statements is true regarding the basicity of aniline and acetanilide?

A. Acetanilide is more basic than aniline.
B. Acetanilide is less basic than aniline.
C. Acetanilide and aniline have the same basicity.
D. The relative basicities cannot be determined.

742. What is the purpose of the first step of the reaction?

A. To increase the solubility of aniline
B. To decrease the solubility of aniline
C. To increase the reactivity of aniline
D. To protect the amine group from reacting with nitric acid

743. Why does the para nitro product form when acetanilide is reacted with nitric acid?

A. Acetanilide contains a para directing group and no ortho or meta directing groups.
B. Acetanilide contains an ortho, para directing group, but it has steric hiderance at the ortho position.
C. Acetanilide contains a meta deactivating group.
D. The nitro group is an ortho, para directing group.

744. If aniline were reacted directly with dilute nitric acid and the amine group were protonated, the resulting ring would be:

A. activated at the meta position
B. activated at the ortho and para positions
C. deactivated and meta directed
D. deactivated and ortho, para directed

Solid derivatives of tertiary amines can be formed from the two reaction shown below. Use these reactions to answer questions 745-749.

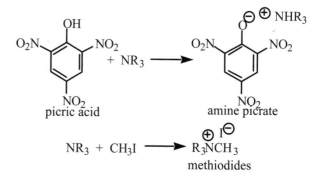

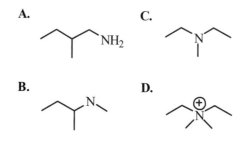

745. The formation of methiodides is a(n):

A. dehydration
B. hydration
C. alkylation
D. dehydrohalogenation

746. The methiodide formed as shown is a(n):

A. imine
B. tertiary amine
C. amide
D. quaternary amine

747. The nitro groups on the picric acid:

A. increase the acidity.
B. decrease the acidity.
C. activate the phenyl ring.
D. are ortho-para directing.

748. Which of the following would readily form an amine picrate?

A. [structure with NH₂ group]
C. [structure with N]
B. [structure with N]
D. [structure with N⁺]

749. The formation of methiodides is a(n):

A. elimination reaction.
B. nucleophilic substitution.
C. aromatic substitution.
D. halogenation.

750. It is common to convert amines that are used as drugs to the hydrochloride salt. An example of this conversion is shown in the reaction below. Why does ephedrine have a foul odor while ephedrine hydrochloride has no odor?

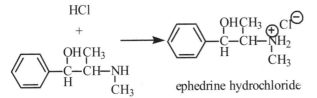

ephedrine hydrochloride

ephedrine

A. The ephedrine hydrochloride has stronger intermolecular forces than ephedrine.
B. The ephedrine hydrochloride has weaker intermolecular forces than ephedrine.
C. Only tertiary amines have foul odors.
D. The ephedrine hydrochloride has a higher molecular weight than ephedrine.

751. In question 750, ephedrine is:

A. a tertiary amine.
B. a quaternary amine.
C. a primary amine.
D. a secondary amine.

Primary and secondary amines can be synthesized by the following reactions. Use these reactions to answer questions 752-756.

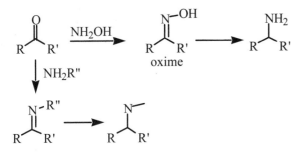

752. The product of the reaction of a primary amine with a ketone is a(n):

A. imine
B. enamine
C. aromatic amine
D. quaternary amine

753. Which of the following reagents would successfully convert an oxime to a primary amine?

A. O_3
B. $LiAlH_4$
C. PBr_3
D. HCN

754. If benzylamine is reacted with cyclopentanone, what would be the product?

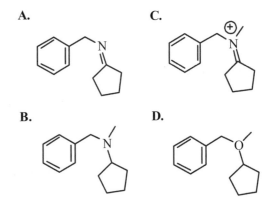

755. Which of the following should be reacted with 2-hexanone to form a primary amine?

A. NH_2OH
B. NH_2CH_3
C. $NH(CH_3)_2$
D. $N(CH_3)_3$

756. Which of the following should be reacted with 2-hexanone to form a secondary amine?

A. NH_2OH
B. NH_2CH_3
C. $NH(CH_3)_2$
D. $N(CH_3)_3$

757. The reaction shown below is a(n):

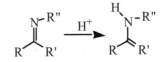

A. oxidation
B. reduction
C. electrophilic addition
D. tautomerization

758. The product in the reaction shown in question 757 is a(n):

A. imine
B. enamine
C. aromatic amine
D. quaternary amine

Use the synthesis below to answer questions 759-762.

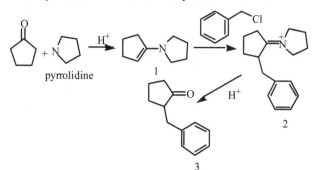

pyrrolidine

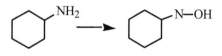

759. Product number 1 of the reaction of cyclopentanone with pyrrolidine is a(n):

   A. imine
   B. enamine
   C. aromatic amine
   D. quaternary amine

760. The conversion of cyclopentanone to the final product is a(n):

   A. substitution
   B. addition
   C. elimination
   D. tautomerization

761. Product number 2 is a(n):

   A. tertiary amine
   B. enamine
   C. aromatic amine
   D. quaternary amine

762. What is the reason for forming product number 1?

   A. Product 1 is more reactive to nuclephilic substitution than an enolate ion.
   B. Product 1 provides stereoselectivity.
   C. Product 1 is more stable than an enolate ion.
   D. Benzyl chloride is not a good nucleophile.

763. In terms of the nitrogen product, the following reaction is a(n):

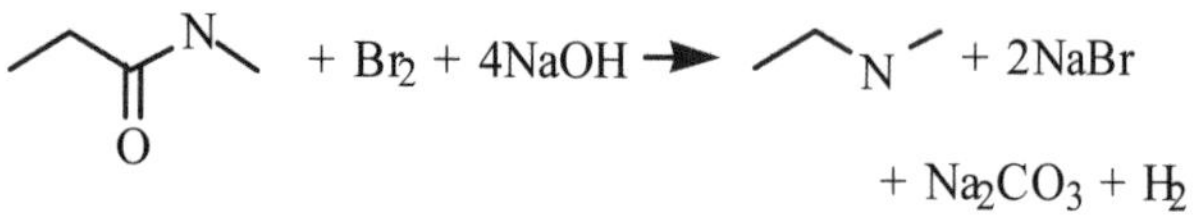

$$\text{(structure)} + Br_2 + 4NaOH \longrightarrow \text{(structure)} + 2NaBr$$
$$+ Na_2CO_3 + H_2$$

   A. reduction
   B. oxidation
   C. dehydration
   D. halogenation

764. Which of the following reagents would achieve the following conversion?

   A. H_2O_2
   B. H_2/Ni
   C. CuCN
   D. PBr_3

765. Which of the following reagents would achieve the following conversion?

   A. H_2O_2
   B. H_2/Ni
   C. CuCN
   D. PBr_3

766. Which of the following reagents would achieve the following conversion?

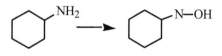

   A. H_2O_2
   B. $LiAlH_4$
   C. CuCN
   D. PBr_3

Fatty Acids and Amino Acids

767. Which of the following is a fatty acid?

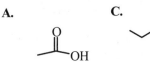

A.

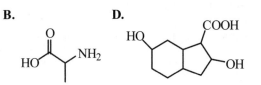

C.

B.

D.

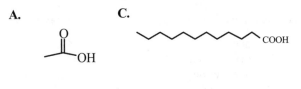

768. Which of the following is an amino acid?

A.

C.

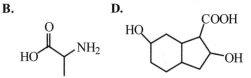

B.

D.

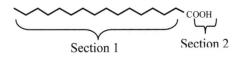

769. Which statement about the fatty acid below is true?

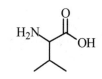

Section 1

Section 2

A. Section 1 is polar and hydrophobic, section 2 is nonpolar and hydrophilic.
B. Section 1 is polar and hydrophilic, section 2 is nonpolar and hydrophobic.
C. Section 1 is nonpolar and hydrophobic, section 2 is polar and hydrophilic.
D. Section 1 is nonpolar and hydrophilic, section 2 is polar and hydrophobic.

770. Which group is the side chain of the amino acid shown below?

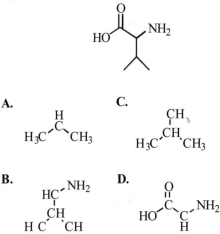

A.

C.

B.

D.

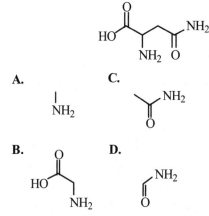

771. Which group is the side chain on the amino acid shown below?

A.

C.

B.

D.

772. The amino acid shown below would be considered:

A. acidic and polar
B. basic and polar
C. polar
D. nonpolar

773. The amino acid shown below would be considered:

A. acidic and polar
B. basic and polar
C. polar
D. nonpolar

774. The amino acid shown below would be considered:

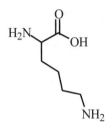

A. acidic and polar
B. basic and polar
C. polar
D. nonpolar

775. The amino acid shown below would be considered:

A. acidic and polar
B. basic and polar
C. polar
D. nonpolar

776. Which of the following is considered the building blocks of proteins?

A. fatty acids
B. triglycerides
C. amino acids
D. sugars

777. Which of the following statements is not true concerning essential amino acids?

A. They cannot be synthesized by the body.√
B. They are needed to make proteins in the human body. √
C. There are eight essential amino acids. √
D. They are beta amino acids.

778. As the chain length of a fatty acid increases,

A. the polarity of the fatty acid decreases and the solubility in water decreases.
B. the polarity of the fatty acid decreases and the solubility in water increases.
C. the polarity of the fatty acid increases and the solubility in water decreases.
D. the polarity of the fatty acid increases and the solubility in water increases.

779. Two amino acids linked by an amide bond is called a:

A. diamine.
B. disaccharide.
C. diglyceride.
D. dipeptide.

780. Two fatty acids linked in by an ester bond is called a:

A. diamine.
B. disaccharide.
C. diglyceride.
D. dipeptide.

Use the reaction shown below to answer questions 781-788.

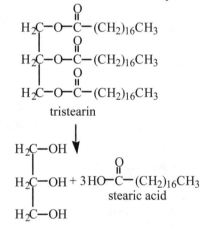

781. Tristearin is a(n):

A. tricarboxylic acid
B. triglycerol
C. triamide
D. triester

782. The conversion of tristearin to stearic acid is a(n):

A. hydrolysis
B. hydrogenation
C. esterification
D. dehydration

783. In the human body, which of the following is most likely used to convert tristearin to stearic acid?

   A. kinases
   B. lipases
   C. NADH
   D. NaOH

784. What reagent is used to convert tristearin into stearate, a soap?

   A. H_2O_2
   B. HCl
   C. NADH
   D. NaOH

785. The reaction described in question 785 is called

   A. lipolysis.
   B. esterification.
   C. oxidation.
   D. saponification.

786. Which of the following is NOT true about stearic acid?

   A. It is amphipathic.
   B. It is amphoteric.
   C. It can be used to form phospholipids.
   D. It can be used to form triacylglycerides.

787. Tristearin is a component of beef fat. Where in the cattle is tristearin stored?

   A. adipose cells
   B. red blood cells
   C. eukaryotic cells
   D. folic cells

788. What initiates the lipolysis of tristearin?

   A. decrease in energy demand
   B. raised levels of carbohydrates
   C. decreased levels of epinephrine
   D. raised levels of epinephrine

789. The structure shown below is a(n):

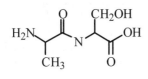

   A. polymer
   B. fatty acid
   C. dipeptide
   D. disaccharide

790. Amino acid sequence determines the secondary structure of proteins. Which of the following best explains the rigidity of protein secondary structure?

   I. hydrogen bonding between amino acids
   II. partial double bond character of the amide linkage
   III. free rotation of the amide linkage

   A. I only
   B. III only
   C. I and II only
   D. I and III only

791. If a solution of valine is at a pH that is equal to the pI for valine, what is the predominant species in the solution?

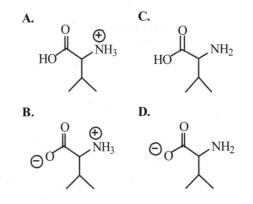

792. If a solution of valine is at a pH that is greater than the pI for valine, what is the predominant species in the solution?

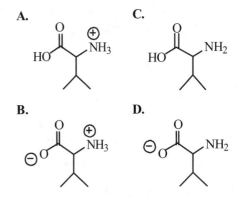

793. If a solution of valine is at a pH that is less than the pI for valine, what is the predominant species in the solution?

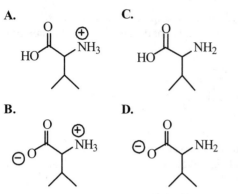

A.

B.

C.

D.

Use the table below to answer questions 794 – 807.

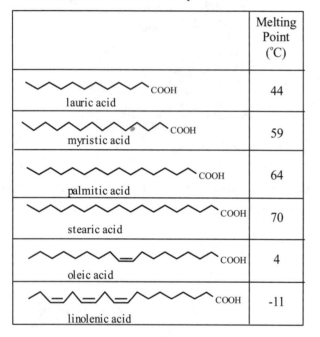

	Melting Point (°C)
lauric acid	44
myristic acid	59
palmitic acid	64
stearic acid	70
oleic acid	4
linolenic acid	-11

794. Which of the following explains why linolenic acid has a much lower melting point than stearic acid?

A. Linolenic acid has a greater molecular weight than stearic acid.
B. Stearic acid has a greater molecular weight than linolenic acid.
C. The *cis* double bonds in linolenic acid make it more difficult for the chains to stack.
D. The *cis* double bonds in linolenic acid make it less difficult for the chains to stack.

795. Which of the following is NOT a saturated fatty acid?

A. oleic acid
B. palmitic acid
C. lauric acid
D. stearic acid

796. The salt of which of the following would be most soluble in water?

A. palmitic acid
B. myristic acid
C. lauric acid
D. stearic acid

797. A fatty acid that is a solid at room temperature is called a fat, while a fatty acid that is a liquid at room temperature is called an oil. Which of the following is an oil?

A. oleic acid
B. palmitic acid
C. lauric acid
D. stearic acid

798. If linoleic acid has the structure shown below, the melting point would be

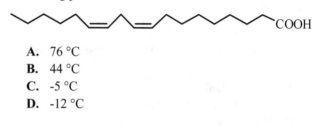

A. 76 °C
B. 44 °C
C. -5 °C
D. -12 °C

799. Arachidic acid contains 20 carbons and no double bonds. What is the melting point of arachidic acid?

A. 76 °C
B. 44 °C
C. -5 °C
D. -12 °C

800. What is the best explanation for the fact that palmitic acid has a greater melting point than lauric acid?

A. Palmitic acid has more surface area than lauric acid.
B. Palmitic acid has less surface area than lauric acid.
C. Palmitic acid has more hydrogen bonding than lauric acid.
D. Palmitic acid has less hydrogen bonding than lauric acid.

801. Lard (solid fat) can be replaced by partially hydrogenated vegetable oil. What is the purpose of hydrogenation?

A. To remove unsaturation and raise the melting point to create a solid
B. To remove unsaturation and lower the melting point to create a solid
C. To remove saturation and raise the melting point to create a solid
D. To remove saturation and lower the melting point to create a solid

802. Which of the following acids would store the greatest amount of energy as a triglyceride?

    A. palmitic acid
    B. myristic acid
    C. lauric acid
    D. stearic acid

803. Which of the following acids would store the least amount of energy as a triglyceride?

    A. palmitic acid
    B. myristic acid
    C. lauric acid
    D. stearic acid

804. The structure of eleostearic acid is shown below. It has a melting point of 49°C. Why is the melting point of eleostearic acid so much higher than linolenic?

    A. Eleostearic acid has more degrees of unsaturation than linolenic acid.
    B. Eleostearic acid has less degrees of unsaturation than linolenic acid.
    C. Eleostearic acid has fewer "kinks" than linolenic acid allowing for better packing.
    D. Eleostearic acid has more "kinks" than linolenic acid allowing for better packing.

805. Given the structure and melting point of eleostearic acid in question 804, why is the melting point of eleosteraric acid greater than oleic acid?

    A. Eleostearic acid has a greater molecular weight than oleic acid.
    B. Eleostearic acid has a smaller molecular weight than oleic acid.
    C. Eleostearic acid has fewer "kinks" than oleic acid allowing for better packing.
    D. The trans bonds found in eleostearic acid form a more rigid structure that is favorable for packing.

806. Micelles are formed when the hydrophobic tails of fatty acids orient themselves to the center of a sphere, leaving the polar carboxylic acid groups on the sphere's surface. Which of the following would form a micelle with the greatest diameter?

    A. palmitic acid
    B. myristic acid
    C. lauric acid
    D. stearic acid

807. Olive oil would have a high percentage by weight of which of the following fatty acids?

A. oleic acid
B. stearic acid
C. palmitic acid
D. myristic acid

808. If glycine is dissolved in an acidic solution that has a pH value below the pI, how many equivalent of NaOH would be required to titrate the glycine to a neutral species?

A. zero equivalents
B. one equivalent
C. two equivalents
D. three equivalents

809. When the pH is equal to the pI of a certain amino acid, a zwitterion forms rather than an uncharged molecule Which of the following must be true concerning the amine group compared to the carboxylate ion?

A. The amine group is more basic than the carboxylate ion of the acid.
B. The amine group is less basic than the carboxylate ion of the acid.
C. The amine group and the carboxylate ion are equally basic.
D. The relative basicities cannot be predicted.

810. Which of the following is true concerning the isoelectric point of arginine?

A. The value is greater than seven.
B. The value is equal to seven.
C. The value is less than seven.
D. The relative value cannot be predicted.

811. Which of the following is true concerning the pI of aspartic acid?

A. pI > 7
B. pI = 7
C. pI < 7
D. pI = 0

812. Ferritin, the iron storage protein, contains two channels. One channel is three-fold and lined with glutamate (the carboxylate ion of glutamic acid) and apartate (the carboxylate ion of apartic acid), while the other is a four-fold channel lined with leucine. Which of the following is true concerning these channels?

A. The three-fold channel is hydrophilic and the four-fold channel is hydrophobic.
B. The three-fold channel is hydrophobic and the four-fold channel is hydrophilic.
C. Both are hydrophilic.
D. Both are hydrophobic.

813. The iron ion leaves ferritin through one of the channels. Given the information in question 812, which channel does the iron most likely use?

A. three-fold channel
B. four-fold channel
C. both the three-fold and the four-fold channels
D. neither the three-fold nor the four-fold channel

814. Which of the following is an amino acid found in the human body?

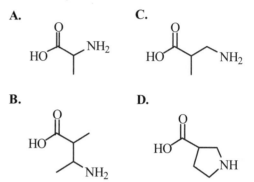

815. Which of the following is an amino acid NOT used to synthesize proteins in the body?

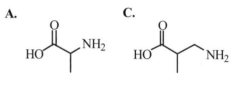

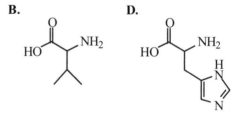

816. A micelle is formed when the carboxylate ions of fatty acids (soap) surround a grease molecule making it soluble in water. Which of the following is a correct representation of a micelle, if the circle represents a grease droplet?

A.

C.

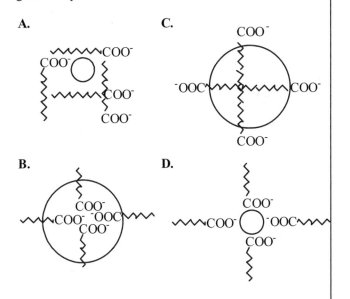

B.

D.

Use the structure of a phosphoglyceride at neutral pH (shown below) to answer questions 817–819.

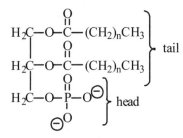

817. The phosphoglyceride would require how many equivalents of HCl to form a neutral molecule?

A. zero equivalents
B. 1 equivalent
C. 2 equivalents
D. 3 equivalents

818. What is true about the phosphoglyceride at neutral pH?

A. The tail of the phosphoglyceride is polar and the head is nonpolar.
B. The tail of the phosphoglyceride is nonpolar and the head is polar.
C. Both the head and tail of the phosphoglyceride are polar.
D. The phosphoglyceride is more polar in acidic solutions.

819. Phosphoglycerides form membrane bilayers in the body. The external surface of the membrane is hydrophilic on both sides and the internal membrane is hydrophobic. What is true about the bilayer membrane?

A. The tails of the phosphoglycerides face the external surfaces and the heads of the phosphoglycerides face the heads of other phosphoglycerides.
B. The heads of the phosphoglycerides face the external surfaces and the tails of the phosphoglycerides face the tails of other phosphoglycerides.
C. The heads and tails of the phosphoglycerides alternate at the external surface.
D. The heads the phosphoglycerides all face the top surface of the membrane and the tails all face the bottom surface.

Use the structure of histidine shown below to answer questions 820-822.

820. Histidine is a(n):

A. polar amino acid.
B. nonpolar amino acid.
C. acidic and polar amino acid.
D. basic and polar amino acid.

821. In the protein hemoglobin, the histidine residue is attached to the Fe of the heme group. The formation of the bond between the histidine and the Fe is a:

A. Lewis acid-base reaction.
B. condensation reaction.
C. dehydration reaction.
D. hydrogenation reaction.

822. What is the pI for histidine?

A. 2.8
B. 3.2
C. 6.9
D. 7.6

Shown below is a titration curve for a 1M solution of glycine titrated with 1M NaOH. Use the curve to answer questions 823-830.

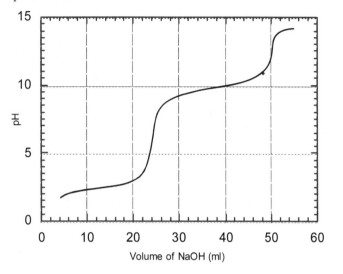

pH

Volume of NaOH (ml)

823. The pI for glycine is:

   A. 2.4
   B. 6.1
   C. 9.8
   D. 12.3

824. According to the titration curve glycine is a(n):

   A. monoprotic acid
   B. diprotic acid
   C. base
   D. diamine

825. What is the pH when the predominant species is the zwitterion?

   A. 2.4
   B. 6.1
   C. 9.8
   D. 12.3

826. What is the pH where 50% of the glycine has a positive charge and 50% is neutral?

   A. 2.4
   B. 6.1
   C. 9.8
   D. 12.3

827. In a solution at pH 12.5, what is the predominant species?

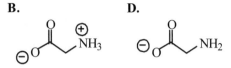

A.   **C.**

B.   **D.**

828. In a solution at pH 6, what is the predominant species?

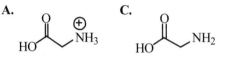

A.   **C.**

B.   **D.**

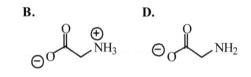

829. In a solution at pH 1.5, what is the predominant species?

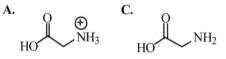

A.   **C.**

B.   **D.**

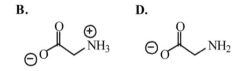

830. When 10 to 20 ml of NaOH are added to 20 ml of NaOH, what hydrogen is neutralized by the NaOH?

   A. the hydrogen on water
   B. the hydrogen on the ammonium group of the amino acid
   C. the hydrogen on the carboxylic acid
   D. No hydrogens are neutralized.

831. The amino acids found in the body are called the L-amino acids because their structure is similar to L-glyceraldehyde. The structure of L-serine is shown below. What is the absolute configuration of L-serine?

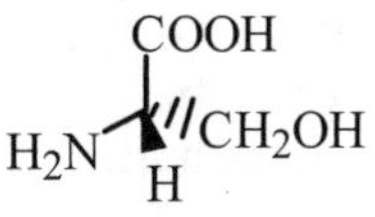

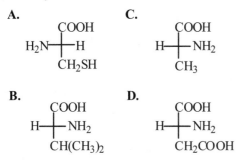

A. R
B. S
C. +
D. –

832. Given the structure of L-serine in question 831, what is the absolute configuration of D-serine?

A. R
B. S
C. +
D. –

833. Which of the following amino acids would be found in the body?

A.

COOH
H₂N——H
CH₂SH

C.

COOH
H——NH₂
CH₃

B.

COOH
H——NH₂
CH(CH₃)₂

D.

COOH
H——NH₂
CH₂COOH

Gelatin can be broken down to its constituent amino acids by refluxing in HCl. Use the reaction shown below to answer questions 834-840.

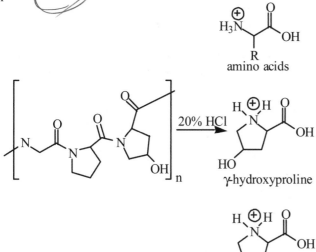

amino acids

20% HCl

γ-hydroxyproline

proline

834. Gelatin is a(n):

A. carbohydrate
B. polypeptide
C. fatty acid
D. steroid

835. Proline is one of the 20 amino acids found in human proteins. It is different from the other 19 because:

A. it is not an alpha amino acid.
B. it is a secondary amine.
C. it contains a nitrogen that can protonated.
D. it is not chiral.

836. Gelatin is formed through the hydrolysis of collagen a structural protein. In this reaction, water is a:

A. catalyst
B. reactant
C. product
D. spectator ion

837. The constituent amino acids can be separated by paper chromatography. One factor that determines how fast the amino acid travels up the paper is the polarity of the amino acid. The more polar amino acids travel slower than the less polar amino acid. Which of the following amino acids will travel up the paper the fastest?

A. cystine (dimer of cysteine)
B. glutamic acid
C. leucine
D. All of the amino acids travel at the same rate.

838. The resulting amino acids from the gelatin hydrolysis have positive charges because:

A. they are zwitterions.
B. the solution has a low pH.
C. the breaking of the amide bond is an oxidation.
D. the breaking of the amide bond is a reduction.

839. Which of the following conversions describes the break down of gelatin?

A. amide to amine and carboxylic acid
B. amine to amide and carboxylic acid
C. carboxylic acid to amine and amide
D. amide to peptide

840. What can be stated about the relative reactivity of the amino acids compared to gelatin?

A. Gelatin is more reactive than the amino acids.
B. The amino acids are more reactive than gelatin.
C. The two have the same reactivity.
D. The relative reactivities cannot be determined.

Use the following table to answer questions 841-851.

Name	Side Chain	pI
glycine	H	6.0
valine	$CH(CH_3)_2$	6.0
serine	CH_2OH	5.7
tyrosine	$CH_2C_6H_4OH$	5.7
asparagine	CH_2CONH_2	5.4
aspartic acid	CH_2COOH	2.8
lysine	$(CH_2)_4NH_2$	9.7
arginine	$(CH_2)_3NHC(NH)NH_2$	10.8

841. If the electrophorectic setup shown below were used to separate a buffer solution of glycine, valine, aspartic acid, and lysine, with a pH 6.0, which amino acid would migrate to the right?

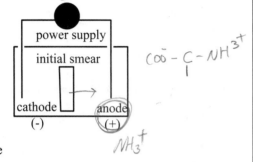

A. glycine
B. valine
C. aspartic acid
D. lysine

842. If the setup shown in question 841 were used to separate a buffer solution of glycine, valine, aspartic acid, and lysine, with a pH 6.0, which amino acid would migrate to the left?

A. glycine
B. valine
C. aspartic acid
D. lysine

843. Which of the following sets of amino acids would be easiest to separate by electrophoresis?

A. glycine, serine, and asparagine
B. valine, serine, and tyrosine
C. aspartic acid, lysine, valine
D. lysine, asparagine, and tyrosine

844. What change would have to be made to the experiment described in question 841 in order to separate lysine, arginine, and valine?

A. raising the pH of the buffer solution to 10.8
B. raising the pH of the buffer solution to 9.7
C. lowering the pH of the buffer solution to 5.7
D. lowering the pH of the buffer solution to 2.8

845. How could the experiment described in question 841 be changed to separate aspartic acid, asparganine, and glycine?

A. Raise the pH of the buffer solution to 9.7
B. Rasie the pH of the buffer solution to 7.2
C. Lower the pH of the buffer solution to 5.4
D. Lower the pH of the buffer solution to 2.4

846. What is the best explanation for the difference in the pI of asparagine and lysine?

A. Asparagine does not contain a nitrogen in the side chain and lysine does contain a nitrogen.
B. Asparagine does contain a nitrogen in the side chain and lysine does not contain a nitrogen.
C. The nitrogen in asapragine's side chain is less basic because it is an amide.
D. The nitrogen in lysine's side chain is less basic because it is an amine.

Amino acids can be synthesized by reductive amination as shown below.

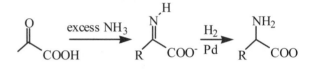

847. Given the information in the table, the reductive amination of the α-ketoacid shown below results in the amino acid:

A. glycine
B. valine
C. aspartic acid
D. lysine

848. Given the information in the table, the reductive amination of the α-ketoacid shown below results in the amino acid:

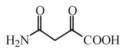

A. glycine
B. serine
C. aspartic acid
D. asparagine

849. Given the information in the table, the reductive amination of which α-ketoacid would result in tyrosine?

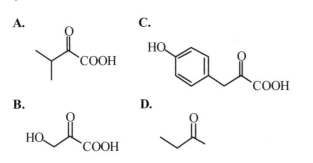

A.

B.

C.

D.

850. If valine is synthesized via reductive amination, what is true about the product?

   A. The absolute configuration is R.
   B. The absolute configuration is S.
   C. The product is racemic.
   D. The product is not chiral.

851. Which of the following statements is NOT true under the conditions of reductive amination?

   A. The intermediate is an imine.
   B. The reduction requires one mole of hydrogen per mole of α-keto carboxylic acid.
   C. The amino acids found in the body could be produced by reductive amination of any keto carboxylic acid.
   D. The reaction produces both R and S amino acids.

852. Methionine has pK_{a1} of 2.28 and a pK_{a2} of 9.21. What is the pI value for methionine?

   A. 2.28
   B. 5.74
   C. 9.21
   D. 10.20

853. The pK_{a1} is the acid dissociation constant for which acidic proton?

   A. protonated nitrogen
   B. carboxylic acid
   C. water
   D. hydroxide ion

Carbohydrates

854. Which of the following is a carbohydrates?

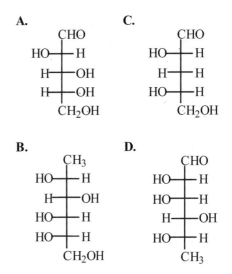

A.

B.

C.

D.

855. Which of the following statements is NOT true regarding carbohydrates?

   A. Carbohydrates have enantiomers.
   B. There are equal numbers of carbons and oxygens.
   C. They are stored in the form of triglycerides.
   D. There are twice the number of hydrogens as carbons.

856. Which of the following statements is NOT true regarding glucose?

   A. Glucose is a ketone.
   B. Glucose is a hexose.
   C. Glucose can exist as a straight chain.
   D. Glucose can exist as a cyclic structure.

857. In living organisms, glucose is oxidized to carbon dioxide and water to produce energy. How many moles of carbon dioxide would result from the complete oxidation of glucose?

   A. 0
   B. 5
   C. 6
   D. 7

858. How many moles of carbon dioxide would result from the complete oxidation of fructose to carbon dioxide and water?

   A. 0
   B. 5
   C. 6
   D. 7

859. Which of the following statements is NOT true about fructose? ✓

A. Fructose is a ketose. ✓
B. Fructose is a hexose. ✓
C. Only L-fructose can be assimilated in the body.
D. There are 8 possible stereoisomers of fructose.

860. Which of the following is L-glucose?

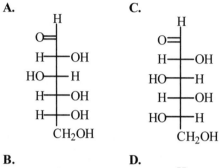

861. Which of the following is L-fructose?

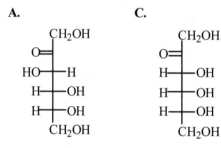

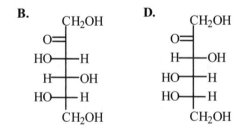

862. Which of the following is D-fructose?

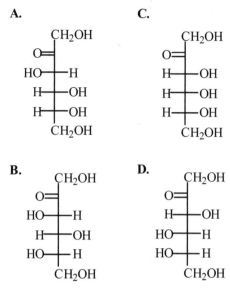

863. What is the relationship between D-glucose and L-glucose?

A. epimers
B. anomers
C. enantiomers
D. structural isomers

864. Which way does L-fructose rotate plane polarized light?

A. clockwise
B. counterclockwise
C. It does not rotate plane polarized light.
D. The direction cannot be determined without measurement.

Use the equilibrium for glucose shown below to answer questions 865 - 870.

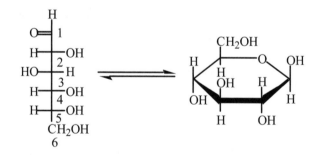

865. The pyranose form of glucose is a(n):

A. acetal
B. hemiacetal
C. ester
D. ether

866. Which chair confirmation corresponds to the pyranose shown at the equilibrium?

A.

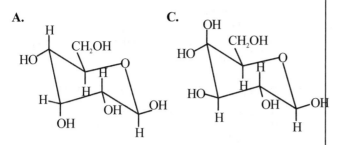

C.

B.

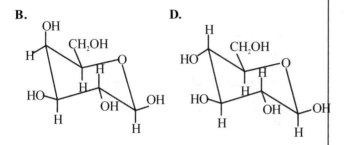

D.

867. The anomeric carbon in the glucose molecule is:

    A. C1
    B. C3
    C. C5
    D. C6

868. The carbon attached to the hydroxyl group that is the nucleophile is:

    A. C1
    B. C3
    C. C5
    D. C6

869. How is the structure shown below related to the pyranose structure in the equilibrium?

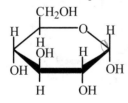

    A. enantiomers
    B. structural isomers
    C. anomers
    D. same molecule

870. How is the structure shown below related to the pyranose structure in the equilibrium?

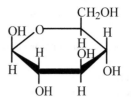

    A. enantiomers
    B. structural isomers
    C. epimers
    D. same molecule

871. Which of the following is D-arabinose?

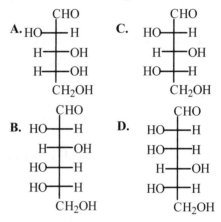

Use the structures of erythrose and threose shown below to answer questions 872-878.

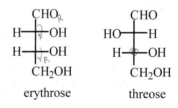

    erythrose   threose

872. What is the relationship between erythrose and threose?

    A. epimers
    B. anomers
    C. enantiomers
    D. structural isomers

873. What is true about the configurations of erythrose and threose?

    A. Erythrose is D and threose is L.
    B. Erythrose is L and threose is D.
    C. Both are D sugars.
    D. Both are L sugars.

874. Which of the following statements is true regarding erythrose and threose?

    A. Erythrose is found in nature and threose is not found in nature.
    B. Erythrose is not found in nature and threose is found in nature.
    C. Both are found in nature.
    D. Neither is found in nature.

875. What is the absolute configuration of carbon 2 and carbon 3 in erythrose, respectively?

    A. R, S
    B. S, R
    C. R, R
    D. S, S

876. What is the absolute configuration of carbon 2 and carbon 3 in threose, respectively?

    A. R, S
    B. S, R
    C. R, R
    D. S, S

877. Which of the following descriptions apply to threose?

    I. aldose
    II. ketose
    III. tetrose

    A. II only
    B. II and III only
    C. I and III only
    D. I, II, and III

878. Degradation of a sugar shortens the chain by one carbon from the top of the Fischer projection releasing CO_2 each time. Which of the following is the result of two successive degradations of D-glucose?

    A. erythrose
    B. threose
    C. fructose
    D. L-glucose

Use the structures of the four aldopentoses that can be formed from D-(+)-glyceraldehyde to answer questions 879-885.

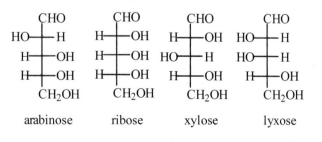

arabinose   ribose   xylose   lyxose

879. D-glyceraldehyde rotates plane polarized light:

    A. clockwise
    B. counterclockwise
    C. It does not rotate plane polarized light.
    D. The direction cannot be determined without measurement.

880. What do the four sugars have in common?

    I. They are aldoses. ✓
    II. They have D configurations.
    III. They rotate light clockwise.

    A. I only
    B. I and II only
    C. II only
    D. I, II, and III

881. How many possible stereoisomers are there for lyxose?

    A. 3
    B. 8
    C. 9
    D. 16

882. The C2 carbon of a sugar will epimerize in the presence of a base. The epimerization of arabinose results in what sugar?

    A. ribose
    B. arabinose
    C. xylose
    D. lyxose

883. The relationship between xylose and ribose is:

    A. enantiomers
    B. structural isomers
    C. epimers
    D. anomers

884. Which carbon shown below is the anomeric carbon?

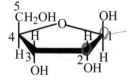

    A. C1
    B. C2
    C. C4
    D. C5

885. The structure shown in questions 884 is the furanose structure for which sugar?

    A. ribose
    B. arabinose
    C. xylose
    D. lyxose

886. The furanose structure shown below is for a(n):

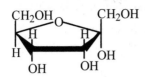

    A. pentanose
    B. aldose
    C. ketose
    D. pyranose

887. Under basic conditions, the carbonyl carbon can shift location by one carbon through enolate and enediol intermediates in a reaction called enediol rearrangement. If D-glucose is dissolved in base and a enediol rearrangement occurs, what is the resulting sugar?

    A. L-glucose
    B. D-glucose
    C. L-fructose
    D. D-fructose

888. The open chain of sugars can be reduced. What would be the product of the reduction of allose?

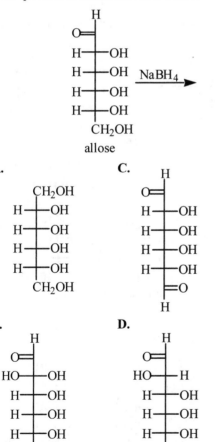

889. What is true about the product from the reduction of allose (question 888)?

    A. The product has no chiral centers.
    B. The product is meso.
    C. The product will rotate plane polarized light.
    D. There is more than one product.

890. If an open-chain fructose is reduced with H_2 and Ni, how many products are produced?

    A. 0
    B. 1
    C. 2
    D. 4

891. Bromine water selectively oxidizes the carbonyl of aldehydes in the presence of alcohols. Which of the following would react with bromine water?

    A. D-glucose
    B. α-D-glucopyranose
    C. D-fructose
    D. α-D-fructofuranose

Use the structure of sucrose shown below to answer questions 892-895.

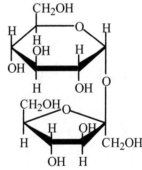

892. Sucrose is composed of what two monosaccharides?

    A. glucose and altose
    B. lactose and glucose
    C. glucose and fructose
    D. fructose and maltose

893. The type of linkage on the pyranose ring is:

    A. α-glycosidic
    B. β-glycosidic
    C. D-glycosidic
    D. L-glycosidic

894. The type of linkage on the furanose ring is:

    A. α-glycosidic
    B. β-glycosidic
    C. D-glycosidic
    D. L-glycosidic

895. Why does sucrose not react with the Tollens' reagent (the test for aldehydes)?

   A. In the open chain form, neither of the monosaccharides are an aldose.
   B. In the open chain form, neither of the monosaccharides are a ketose.
   C. In sucrose, the anomeric carbons are part of an acetal and a ketal which are less reactive to hydrolysis than a hemiacetal and hemiketal.
   D. In sucrose, the anomeric carbons are hemiacetals which are less reactive to hydrolysis than acetals.

Use the structure of cellulose shown below to answer questions 896 - 898.

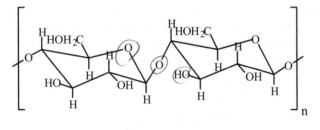

896. Cellulose is a:

   A. carbohydrate
   B. polypeptide
   C. fatty acid
   D. steroid

897. The type of linkages between the monosaccharides in cellulose are:

   A. α-glycosidic
   B. β-glycosidic
   C. D-glycosidic
   D. L-glycosidic

898. The anomeric carbon is a(n):

   A. acetal
   B. hemiacetal
   C. ester
   D. ether

The basic carbohydrate portion of the structure for a ribonucleoside is shown below. Use this structure to answer questions 899-900.

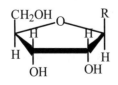

899. Given that R is a pyrimidine base, the carbohydrate portion of the ribonucleoside is a:

   A. pyranose of a ketose.
   B. pyranose of an aldose.
   C. furanose of a ketose.
   D. furanose of an aldose.

900. If the pyrimidine base is replaced by an OH group, what would be the open-chain structure?

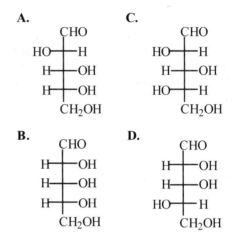

Use the synthesis of tartaric acid shown below to answer questions 901-906.

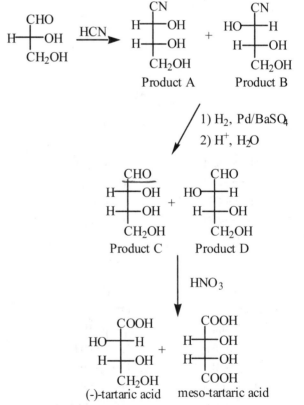

901. What is the relationship between Products A and B?

    A. enantiomers
    B. anomers
    C. structural isomers
    D. epimers

902. The formation of tartaric acid from Products C and D is a(n):

    A. reduction
    B. oxidation
    C. epimerization
    D. hydrolysis

903. How will Product D rotate plane polarized light?

    A. clockwise
    B. counter clockwise
    C. The direction cannot be determined.
    D. Product D will not rotate light.

904. How will meso-tartaric acid rotate plane polarized light?

    A. clockwise
    B. counter clockwise
    C. The direction cannot be determined.
    D. Meso-tartaric acid will not rotate light.

905. Product C is a:

    A. D-aldotetrose
    B. L-aldotetrose
    C. D-ketotetrose
    D. L-ketotetrose

906. Product C will form:

    A. a pyranose ring
    B. a furanose ring
    C. cyclohexyl ring
    D. cyclopentyl ring

Use the structure of lactose shown below to answer questions 907-910.

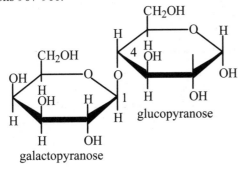

907. Lactose is a disaccharide with a 1,4 glycosidic linkage. The anomeric carbon on the glucopyranose is

    A. C1
    B. C2
    C. C3
    D. C4

908. The linkage in lactose is a(n):

    A. β-galactopyranosyl
    B. α-galactopyranosyl
    C. α-glucopyranosyl
    D. β-glucopyranosyl

909. Which is the open-chain structure of galactose?

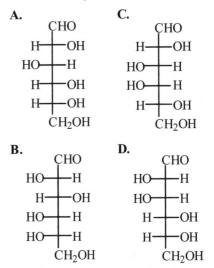

910. What is the relationship between galactose and glucose?

    A. enantiomers
    B. anomers
    C. diastereomers
    D. epimers

Lab Techniques

911. In what units are the chemical shifts of hydrogens in a nmr spectrum measured?

    A. wavenumbers
    B. ppm
    C. wavelengths
    D. absorbance units

912. NMR stand for:

    A. net magnetic response
    B. non-magnetic response
    C. nuclear magnetic response
    D. nuclear magnetic resonance

913. Which technique separates compounds based on boiling point?

    A. chromatography
    B. distillation
    C. crystallization
    D. extraction

914. Which technique separates compounds based on polarity and affinity for silica gel?

    A. chromatography
    B. distillation
    C. crystallization
    D. extraction

915. Which technique separates compounds based on solubility in organic and aqueous solvents?

    A. chromatography
    B. distillation
    C. crystallization
    D. extraction

916. The energy absorbed by a molecule to produce an IR spectrum is:

    A. infrared
    B. microwave
    C. magnetic
    D. radio wave

917. An R_f value is:

    A. the ratio of two distances.
    B. the strength of a magnetic field.
    C. the ratio of hydrogens.
    D. the frequency of a bond vibration.

918. The chemical shift of a peak in a NMR spectrum indicates:

    A. the electronic environment.
    B. the number of neighboring hydrogens.
    C. the relative number of hydrogens.
    D. the frequency of the bond vibration.

919. The splitting of a peak in a NMR spectrum indicates:

    A. the electronic environment.
    B. the number of neighboring hydrogens.
    C. the relative number of hydrogens.
    D. the frequency of the bond vibration.

920. What are the units of IR peaks?

    A. ppm
    B. Hz
    C. cm^{-1}
    D. δ

921. Which of the following is a possible R_f value for an unknown compound?

    A. 0.4
    B. 1.2
    C. 1.6
    D. 2.0

922. What is true about a compound that has an R_f value of 0 in a typical thin layer chromatography experiment?

    A. The compound moved at the solvent front, and it is not very polar.
    B. The compound moved at the solvent front, and it is very polar.
    C. The compound did not move from the starting line, and it is not very polar.
    D. The compound did not move from the starting line, and it is very polar.

923. What is true about a compound that has an R_f value of 1 in a typical thin layer chromatography experiment?

    A. The compound moved at the solvent front, and it is not very polar.
    B. The compound moved at the solvent front, and it is very polar.
    C. The compound did not move from the starting line, and it is not very polar.
    D. The compound did not move from the starting line, and it is very polar.

924. The area under a peak in a NMR spectrum indicates:

    A. the electronic environment.
    B. the number of neighboring hydrogens.
    C. the relative number of hydrogens.
    D. the frequency of the bond vibration.

925. Why is the range of absorption 2100-2260 for a triple bond when it is 1620-1680 for a double bond?

    A. Double bonds are stiffer than triple bonds.
    B. Triple bonds are stiffer than double bonds.
    C. Triple bonds have fewer pi bonds.
    D. Double bonds have less p character in the hybridized orbital.

926. What causes a peak in a NMR spectrum?

    A. A deflection of a molecule by a magnetic field.
    B. The absorption of electromagnetic radiation.
    C. The reorientation of a molecule in a magnetic field.
    D. A reflection of a frequency of electromagnetic radiation.

927. How many peaks will appear in an NMR spectrum of the molecule shown below?

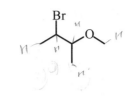

    A. 4
    B. 5
    C. 10
    D. 12

928. How many peaks that are not split will be found in the NMR spectrum of the compound shown in question 927?

    A. 0
    B. 1
    C. 2
    D. 3

929. What change occurs in a molecule when IR radiation is absorbed?

    A. The spin of the electrons flips.
    B. The vibrations of the bonds change.
    C. The rotation around the bond axes change.
    D. The translational motion changes.

930. What change occurs in a molecule when energy is absorbed to produce an NMR spectrum?

    A. The spin of the nucleus flips.
    B. The vibrations of the bonds change.
    C. The rotation around the bond axes change.
    D. The translational motion changes.

931. A stretch at 3400 cm^{-1} in an IR spectrum indicates the presence of what functional group?

    A. ketone
    B. alcohol
    C. alkene
    D. alkyne

932. A stretch at 1700 cm^{-1} in an IR spectrum indicates the presence of what functional group?

    A. ketone
    B. alcohol
    C. alkene
    D. alkyne

Use the structure shown below to answer questions 933–937.

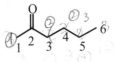

933. How many peaks will appear in a NMR spectrum 2-hexanone?

    A. 5
    B. 6
    C. 11
    D. 12

934. What is true about the peak that is produced by the hydrogens on carbon 1 in a nmr spectrum?

    A. The peak will integrate as 2 and will split into a doublet.
    B. The peak will integrate as 2 and will split into a triplet.
    C. The peak will integrate as 1 and will be split into a triplet.
    D. The peak will integrate as 3 and will be a singlet.

935. What is true about the peak that is produced by the hydrogens on carbon 3 in a nmr spectrum?

    A. The peak will integrate as 2 and will split into a doublet.
    B. The peak will integrate as 2 and will split into a triplet.
    C. The peak will integrate as 1 and will be split into a triplet.
    D. The peak will integrate as 3 and will be a singlet.

936. The peak that is farthest upfield in the nmr spectrum of 2-hexanone is a triplet. To which carbon are the hydrogens responsible for the peak attached?

    A. carbon 1
    B. carbon 2
    C. carbon 5
    D. carbon 6

937. Which of the following peaks will be present in an IR spectrum of 2-hexanone?

    A. 1680 cm^{-1}
    B. 2220 cm^{-2}
    C. 3200 cm^{-2}
    D. 3500 cm^{-2}

Use the table of physical constants below to answer questions 938–943.

Alcohol	Molecular Weight	Boiling point
benzyl alcohol	108.14	205 °C
cyclohexyl alcohol	100.16	161 °C
isoamyl alcohol	88.15	130 °C
octyl alcohol	130.23	196 °C
propyl alcohol	60.10	82 °C

938. If a mixture of the alcohols shown in the table is distilled, the first drops to roll down the condenser would contain which alcohol?

   A. benzyl alcohol
   B. isoamyl alcohol
   C. octyl alcohol
   D. propyl alcohol

939. If a mixture of the alcohols shown in the table is distilled, which alcohol would be the last to come across?

   A. benzyl alcohol
   B. isoamyl alcohol
   C. octyl alcohol
   D. propyl alcohol

940. Which of the following mixtures would be the easiest to separate using distillation?

   A. benzyl alcohol and propyl alcohol
   B. isoamyl alcohol and propyl alcohol
   C. benzyl alcohol and octyl alcohol
   D. octyl alcohol and cyclohexyl alcohol

941. Which of the following mixtures would be the hardest to separate using distillation?

   A. benzyl alcohol and propyl alcohol
   B. isoamyl alcohol and propyl alcohol
   C. benzyl alcohol and octyl alcohol
   D. octyl alcohol and cyclohexyl alcohol

942. Benzyl acetate (boiling point 206°C) is synthesized by reacting acetic anhydride with benzyl alcohol. What is true about the distillation to separate the ester from the alcohol starting material?

   A. The ester and alcohol will come across as pure vapor.
   B. The distillation cannot be carried out with a simple distillation apparatus.
   C. The benzyl acetate will come across first.
   D. The distillation will be fast.

943. Propyl acetate (boiling point 102°C) is synthesized by reacting acetic anhydride with propyl alcohol. What is true about the distillation to separate the ester from the alcohol starting material?

   A. The ester and alcohol will be easy to separate.
   B. The distillation cannot be carried out with a simple distillation apparatus.
   C. The propyl acetate will come across first.
   D. The distillation will be fast.

Use the structure of morpholine shown below to answer questions 944-946.

944. How many peaks will appear in a NMR spectrum of morpholine?

   A. 2
   B. 3
   C. 4
   D. 6

945. In the NMR spectrum of morpholine, how many triplet peaks are present?

   A. 1
   B. 2
   C. 3
   D. 6

946. Which of the following peaks will be present in an IR spectrum of morpholine?

   A. 1680 cm^{-1}
   B. 1720 cm^{-2}
   C. 2000 cm^{-2}
   D. 3400 cm^{-2}

Refer to the scheme below to answer questions 947-952.

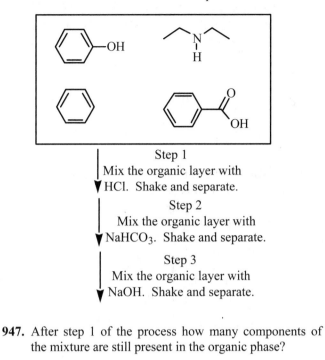

Step 1
Mix the organic layer with
HCl. Shake and separate.

Step 2
Mix the organic layer with
NaHCO$_3$. Shake and separate.

Step 3
Mix the organic layer with
NaOH. Shake and separate.

947. After step 1 of the process how many components of the mixture are still present in the organic phase?

   A. 1
   B. 2
   C. 3
   D. 4

948. What is present in the aqueous phase at the end of step 1?

   A. phenol
   B. diethylamine
   C. benzoic acid
   D. benzene

949. What is present in the organic phase at the end of step 3?

   A. phenol
   B. diethylamine
   C. benzoic acid
   D. benzene

950. What is present in the aqueous phase at the end of step 2?

   A. phenol
   B. diethylamine
   C. benzoic acid
   D. benzene

951. If step 1 and step 2 are reversed, how will the separations change?

   A. Phenol and benzoic acid will not be separated.
   B. Phenol and diethylamine will not be separted.
   C. Benzene and benzoic acid will not be separted.
   D. All four components will still be separated.

952. If step 2 and step 3 are reversed, how will the separations change?

   A. Phenol and benzoic acid will not be separated.
   B. Phenol and diethylamine will not be separated.
   C. Benzene and benzoic acid will not be separated.
   D. All four components will still be separated.

Refer to the nmr spectrum shown below to answer questions 953–963.

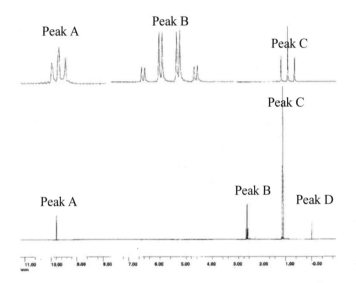

953. What is the purpose of the peak D?

   A. to reference the integration
   B. to reference the chemical shift
   C. to reference the splitting
   D. to reference the coupling

954. How many hydrogens that are not chemically equivalent does peak A have on the neighboring carbon?

   A. 1
   B. 2
   C. 3
   D. 4

955. What is true about peak B compared to peak A?

    A. Peak A is downfield from peak B.
    B. Peak A is upfield from peak B.
    C. Peak A has a greater field strength than peak B.
    D. Peak A and peak B are attached to the same carbon.

956. This spectrum is of a(n):

    A. ketone
    B. aldehyde
    C. ester
    D. amine

957. Which peak is the farthest upfield?

    A. Peak A
    B. Peak B
    C. Peak C
    D. Peak D

958. Which peak results from the proton that has the least electron shielding?

    A. Peak A
    B. Peak B
    C. Peak C
    D. Peak D

959. Which peak has the most hydrogens attached to the neighboring carbon?

    A. Peak A
    B. Peak B
    C. Peak C
    D. Peak D

960. Peak B is a:

    A. triplet of doublets
    B. quartet of doublets
    C. triplet
    D. singlet

961. What is the δ of peak A?

    A. 1.2 ppm
    B. 3.0 ppm
    C. 4.0 ppm
    D. 9.8 ppm

962. Given that peak D is a reference peak, how many different types (in terms of chemical equivalence) of hydrogens are there in the compound used to obtain the spectrum?

    A. 1
    B. 3
    C. 10
    D. 14

963. Which of the following compounds corresponds to the spectrum?

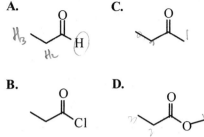

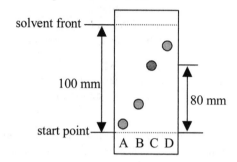

Use the drawing of the thin layer chromatography (tlc) plate below to answer question 964 - 968.

964. Which compound is the most polar?

    A. the compound in lane A
    B. the compound in lane B
    C. the compound in lane C
    D. the compound in lane D

965. Which compound is the least polar?

    A. the compound in lane A
    B. the compound in lane B
    C. the compound in lane C
    D. the compound in lane D

966. Which compound has the greatest R_f value?

    A. the compound in lane A
    B. the compound in lane B
    C. the compound in lane C
    D. the compound in lane D

967. Which compound has the smallest R_f value?

    A. the compound in lane A
    B. the compound in lane B
    C. the compound in lane C
    D. the compound in lane D

968. What is the R_f value for compound C?

    A. 0.8
    B. 1.25
    C. 80
    D. 100

Refer to the proton nmr spectrum shown below to answer questions 969–981.

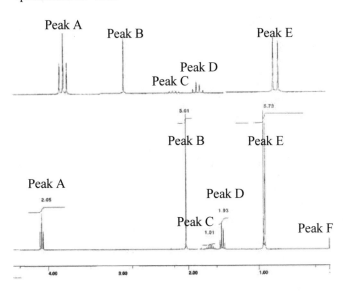

969. Given that peak F is a reference peak, how many types of hydrogens that are not chemically equivalent does the compound have?

A. 5
B. 6
C. 14
D. 15

970. What is the chemical shift of peak E?

A. 0.9
B. 1.2
C. 2
D. 5.73

971. Which peak is farthest downfield?

A. Peak A
B. Peak B
C. Peak E
D. Peak F

972. Which peak corresponds to the type of hydrogen that has the most hydrogens that are not chemically equivalent on neighboring carbons?

A. Peak A
B. Peak C
C. Peak D
D. Peak E

973. Which peak corresponds to the most hydrogens on the molecule?

A. Peak A
B. Peak B
C. Peak C
D. Peak E

974. Peak A is a:

A. singlet
B. doublet
C. triplet
D. quartet

975. How many relative hydrogens produced peak A?

A. 1
B. 2
C. 3
D. 4

976. How many hydrogens are on the neighboring carbon of the protons that produced peak A?

A. 1
B. 2
C. 3
D. 4

977. The 5.73 above peak E indicates:

A. the area under the peak
B. the number of corresponding hydrogens
C. the number of neighboring hydrogens
D. the chemical shift

978. Which peak is produced by the hydrogen(s) closest to an electronegative atom?

A. Peak A
B. Peak C
C. Peak D
D. Peak E

Use the structure shown below to answer questions 979–981.

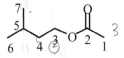

979. Given that the structure shown above is the molecule that produced the NMR spectrum, which peak corresponds to the hydrogens attached to carbon 3?

A. Peak A
B. Peak B
C. Peak C
D. Peak E

980. Peak C in the spectrum corresponds to the hydrogen(s) attached to which carbon?

    A. the carbon labeled 1
    B. the carbon labeled 4
    C. the carbon labeled 5
    D. the carbon labeled 6

981. Peak E in the spectrum corresponds to the hydrogens attached to which carbons?

    I. The carbon labeled 1
    II. The carbon labeled 6
    III. The carbon labeled 7

    A. I only
    B. II only
    C. II and III
    D. I, II, and III

Five solutions were spotted on a tlc plate. These solutions contained caffeine (lane A), aspirin (lane B), acetaminophen (lane C), Excedrin (lane D), and Tylenol (lane E). Use the resulting tlc plate shown below to answer questions 982-987.

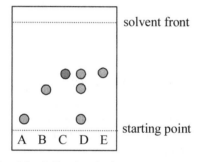

982. Which of the following is the most polar?

    A. caffeine
    B. aspirin
    C. acetaminophen
    D. Tylenol

983. Which of the following has the greatest R_f value?

    A. caffeine
    B. aspirin
    C. acetaminophen
    D. all have the same R_f value

984. What is the R_f of caffeine?

    A. 0
    B. 0.1
    C. 0.8
    D. 1.0

985. Which of the following pain relief ingredients are contained in Excedrin?

    I. caffeine
    II. acetaminophen
    III. aspirin

    A. I only
    B. II only
    C. II and III
    D. I, II, and III

986. Which of the following pain relief ingredients are contained in Tylenol?

    I. caffeine
    II. acetaminophen
    III. aspirin

    A. I only
    B. II only
    C. II and III
    D. I, II, and III

987. The R_f value of a substance can be changed by altering the ratio of solvents in the developing solution. If the developing solution for the tlc was a solution of 5% methanol in ethyl acetate, what could be done to increase the R_f value of caffeine?

    A. Increase the amount of methanol.
    B. Decrease the amount of methanol.
    C. Run the chromatogram for a long time period.
    D. The R_f value is constant.

988. Water and ethanol form an azeotrope of 96% ethanol and 4% water at a boiling point of 78°C. What is true about ethanol that is purified by distillation?

    A. It will contain more ethanol than water.
    B. It will contain more water than ethanol.
    C. It will contain 100% ethanol.
    D. It will contain 100% water.

989. Acetanilide can be purified by crystallization from water. If acetanilide has a solubility of 1 g / 185 ml in water at room temperature and 1 g/ 20 ml in hot water, how many grams should be dissolved in 100 ml of hot water at the start of the crystallization?

    A. 1 gram
    B. 3 grams
    C. 5 grams
    D. 7 grams

990. If the crystallization of acetanilide is carried out with top efficiency, what is true about the dried crystallized product?

    A. All of the acetanilide will be recovered.
    B. Some of the acetanilide will remain in the water.
    C. The acetanilide has a lower purity after crystallization.
    D. The crystallized product will weigh more than the original.

Refer to the reaction scheme below to answer questions 991-994.

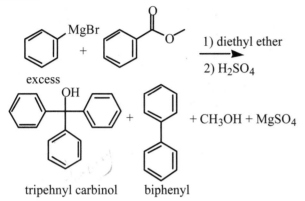

excess

tripehnyl carbinol   biphenyl

991. When the synthesis of triphenyl carbinol is completed, the reaction mixture is in an acidic aqueous solution. This solution is extracted with diethyl ether. Which of the following products is present in the aqueous solution after the extraction is complete?

    I. $MgSO_4$
    II. biphenyl
    III. triphenyl carbinol

    A. I only
    B. III only
    C. II and III only
    D. I, II, and III

992. Given the information in question 991, which of the following products is present in the organic solution after the extraction is complete?

    I. $MgSO_4$
    II. biphenyl
    III. triphenyl carbinol

    A. I only
    B. III only
    C. II and III only
    D. I, II, and III

993. Diethyl ether has a density of 0.715. What is true about the extraction?

    A. The organic layer is on the top.
    B. The aqueous layer is on the top.
    C. The solutions are miscible.
    D. The boiling point of the organic layer is depressed.

994. Given the solubilities in the table below, which solvent would be the best choice to crystallize triphenyl carbinol in order to separate it from biphenyl?

	Insoluble	Soluble
triphenyl carboninol	water, hexane	ether, benzene
biphenyl	water	ether

    A. ether and benzene
    B. hexane and water
    C. ether and hexane
    D. water

Caffeine is isolated from the other components found in tea by extraction with sodium hydroxide. Refer to the structures of the tea components below to answer questions 995 - 1001.

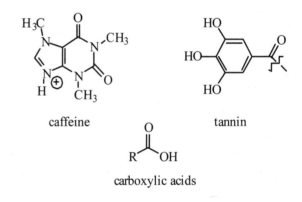

caffeine   tannin

carboxylic acids

995. After the addition of sodium hydroxide to the tea solution and extraction with dichloromethane, what components are present in the organic layer?

    I. caffeine
    II. carboxylic acids
    III. tannins

    A. I only
    B. II only
    C. II and III
    D. I, II, and III

996. After the addition of sodium hydroxide to the tea solution and extraction with dichloromethane, what components are present in the aqueous layer?

   I. caffeine
  II. carboxylic acids
 III. tannins

   A. I only
   B. II only
   C. II and III only
   D. I, II, and III

997. If the tea solution were extracted without the sodium hydroxide present, what components would be present in the aqueous phase after the extraction?

   I. caffeine
  II. carboxylic acids
 III. tannins

   A. I only
   B. II only
   C. II and III only
   D. I, II, and III

998. If the tea solution were extracted without the sodium hydroxide present, what components would be present in the organic phase after the extraction?

   I. caffeine
  II. carboxylic acids
 III. tannins

   A. I only
   B. II only
   C. II and III
   D. I, II, and III

999. If the tea solution were extracted with sodium bicarbonate instead of sodium hydroxide, what components would be present in the organic phase after the extraction?

   I. caffeine
  II. carboxylic acids
 III. tannins

   A. I only
   B. II only
   C. I and III only
   D. I, II, and III

1000. If the tea solution were extracted with sodium bicarbonate instead of sodium hydroxide, what components would be present in the aqueous phase after the extraction?

   I. caffeine
  II. carboxylic acids
 III. tannins

   A. I only
   B. II only
   C. I and III
   D. I, II, and III

1001. Given that dichloromethane has a density of 1.362, what is true about the extraction?

   A. The organic layer is on the top.
   B. The aqueous layer is on the top.
   C. The solutions are miscible.
   D. The boiling point of the organic layer is depressed.

1. **A** A carbon makes four bonds to complete its octet. Each pair of electrons represents one bond. In choices B and D, the second carbon makes five bonds. In choice C the third carbon makes only three bonds.
2. **D** Answers A and B both contain too many electrons [1(H) + 4(C) + 5(N) = 10 electrons]. Answer C has the correct number of electrons, however, there is a formal charge of –2 on the carbon and +2 on the nitrogen. Answer D has 10 electrons and formal charges of zero on all atoms making it the correct answer.
3. **B** NH_3 is the only answer with a lone pair of electrons. Answers C and D are both well-known Lewis acids. They have empty orbitals to receive electrons.
4. **B** Answers A, C, and D all have lone pairs of electrons. Answer B is an ammonium ion so the lone pair of electrons found on ammonia is now used to form the bond with the fourth hydrogen.
5. **B** Newman projections are drawn for a given carbon-carbon bond. If the molecule is rotated so that the eye is looking down the axis of the bond, the molecule in front is drawn as the intersection of three lines, the obscured carbon is drawn as a circle.
6. **C** See answer to question 5.
7. **B** The methyl group is attached to the front carbon and the hydroxide group is attached to the back carbon.
8. **C** The Fischer projection shown is a shorthand representation of the structure shown below. In a Fischer projection, all horizontal lines are oriented out of the page, while vertical lines are oriented into the page. The hydroxide groups are coming out of the page.

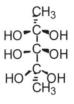

9. **D** The Fischer projection shown is a shorthand representation of the structure shown above. The methyl groups are going into the page.
10. **B** The intersection of two lines always represents a carbon atom in a Fischer projection.
11. **A** In the dash-line-wedge formula, the dark wedge symbolizes the atom(s) that comes out of the page.
12. **B** In the dash-line-wedge formula, the dashed wedge symbolizes the atom(s) that goes into of the page.
13. **C** In the dash-line-wedge formula, the lines symbolize the atom(s) that are in the plane of the page.
14. **B** The amide functional group on the molecule below is circled. The "am" prefix indicates a functional group with a nitrogen atom.

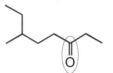

15. **C** The ketone functional group is circled in the compound below.

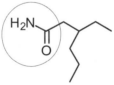

16. **B** The ester functional group has one carbonyl connected to an oxygen atom that has a carbon chain attached.
17. **A** The ether functional group has an oxygen atom with two carbon chains attached.
18. **B** The ester group in isoamylacetate is circled below.

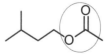

19. **C** The question states that THF is an ether and answer C is the only ether. Answer A is an amide, B is an ester and carboxylic acid and D is an alcohol.
20. **B** The functional groups are indicated below.

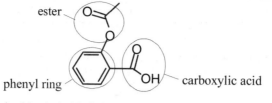

ester

phenyl ring — carboxylic acid

21. B The single amine group on luminol is circled below.

amine

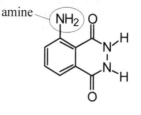

22. C The two amide groups are circled below.

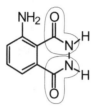

23. D Menthol is a cycloalkane with an alcohol.
24. B Menthol has a hydroxyl group and limonene has a double bond.
25. A The functional groups of eugenol are circled below.

aromatic ring

hydroxyl

alkene

ether

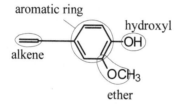

26. D The only difference between the two structures is between the alcohol and ester.
27. D In the equation below n is the number of carbons and x is the number of hydrogens.

$$\frac{(2n+2)-x}{2} = \frac{(2(8)+2)-12}{2} = 3$$

28. D When calculating the index of hydrogen deficiency, O atoms are not included in the calculation.

$$\frac{(2n+2)-x}{2} = \frac{(2(8)+2)-12}{2} = 3$$

29. D When calculating the index of hydrogen deficiency, a halogen counts as a hydrogen.

$$\frac{(2n+2)-x}{2} = \frac{(2(8)+2)-12}{2} = 3$$

30. B Each ring in a structure corresponds to a hydrogen deficiency. If the carbons and hydrogens in the structure are counted, the formula is C_9H_{16} which equals an index of 2 when used in the formula.

$$\frac{(2n+2)-x}{2} = \frac{(2(9)+2)-16}{2} = 2$$

31. B Each double bond in a structure corresponds to a hydrogen deficiency. If the carbons and hydrogens in the structure are counted and the oxygen atom is ignored, the formula C_5H_8O which equals an index of 2 when used in the formula.

$$\frac{(2n+2)-x}{2} = \frac{(2(5)+2)-8}{2} = 2$$

32. C With two rings and a double bond there is a hydrogen deficiency index of 3. If the carbons and hydrogens in the structure are counted, the molecular formula is $C_{10}H_{15}Br$ which equals an index of 3.

$$\frac{(2n+2)-x}{2} = \frac{(2(10)+2)-16}{2} = 3$$

33. C All the nitrogens in the Calcofluor are contained in amine groups.

34. D A phenyl group has 4 degrees of unsaturation (i.e., a hydrogen deficiency index of 4), one ring and three double bonds. The formula for a phenyl ring is C_6H_6.

35. C The functional groups on the structure are identified below.

36. D The longest chain is seven carbons indicating a heptane (eliminating answers A and B) and the tert-butyl group is on the lowest number carbon making D the right answer.

37. B The longest chain is eight carbons indicating octane (eliminating answers C and D). The alcohol substituent needs to have the lowest number making B the right answer.

38. D The IUPAC name indicates that there are two substituents on the cyclooctane ring eliminating answers A and B. One of the substituents is butyl (four carbons) eliminating C.

39. A The question indicates that the product is an ether. Only answer A is an ether. Answer B is an alkyne, C an alkene, and D is an aldehyde.

40. B The question indicates that an anhydride is the best choice, and answer C is the only anhydride. Answer A is a carboxylic acid, B an alcohol, and D an ester.

41. B The amide groups are circled in the structure below.

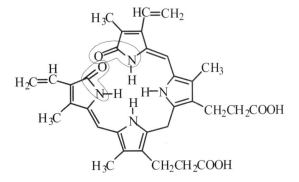

42. B The amine groups are circled in the structure below.

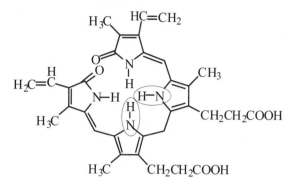

43. B The carboxylic acid groups are circled in the structure below.

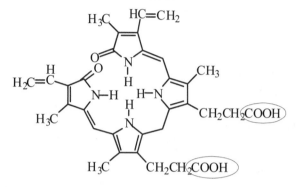

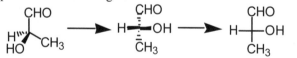

44. C The question states that 2,4-DNP reacts with ketones and aldehydes. Answer C is a ketone. Answer A is an alcohol, B an amine, and D is an ether.

45. B Question 44 indicates that 2,4-DNP reacts with ketones and aldehydes, therefore both acetone (a ketone) and heptanal (an aldehyde) would react with the reagent. Only ethanol, an alcohol, will not react with the reagent, making B the correct answer.

46. D Since the reaction starts with cyclopentene, answers A and C are ruled out as products because hydration does not add carbons. The question states that the product is an alcohol, so answer B is ruled out (it is the reagent), leaving D as the right answer.

47. C The question states that the reaction replaces the halogen with an amine group. Answer C is the only answer that replaces the 1-chloro group with an amine group. Answer D is the reagent.

48. B The transformation is two hydrogens replaced by a carbonyl making the transformation an alkane to a ketone (cylco because it takes place on a cycloalkane).

49. C According to the IUPAC name, carbon number 1 is the carbon attached to the chloride atom. Counting the carbons, the third carbon is the intersection of the three lines and the circle is carbon number 4.

50. A The question asks for the Fischer projection so answers B and D are ruled out. If the molecule is rotated as shown below. The hydroxide group is oriented to the right.

51. A The question asks for dash-line-wedge formula so answers B and D are ruled out. Following the rotation below answer A is correct. If you have trouble seeing this rotation, build a model.

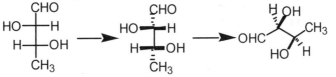

52. D The bonds between atoms result from the attraction of electrons to the protons in the nuclei of the atoms. Because the attraction involves charges, it is electrostatic.

53. C There is one sigma bond between the carbon and the nitrogen, therefore one electron from each atom contributes to the bond for a total of 2 electrons.

54. B Only one sigma bond can form between two atoms, therefore, the remaining bonds must be pi. There are a total of three bonds in HCN, so there must be one sigma and 2 pi.

55. B There are two pi bonds between the carbon and the nitrogen atoms. The nitrogen contributes one electron to each pi bond for a total of two.

56. A There are two p orbitals on each atom forming the two pi bonds, leaving one p and one s orbital to hybridize to form the sigma bond. The resulting hybridization is sp. (Triple bonds are always sp hybridized.)

57. A The formula for formal charge is:

> Formal charge = group number of the atom – number of electrons in lone pairs – ½(number of electrons in bonding pairs)

> In this case, the group number of nitrogen is 5. There are 2 electron in lone pairs and 6 in bonding pairs: Formal charge = $5 - 2 - \frac{1}{2}(6) = 0$

58. A Sigma bonds are the most stable bonds between atoms. They are the last to break in reactions.

59. C An alkyne is a triple bond which involves one sigma and two pi bonds (ruling out answers A and B). The p orbitals that form the two pi bonds are oriented perpendicularly to each.

60. B The table shows that a carbon-carbon single bond (a sigma bond) is 83 kcal.

61. A The table demonstrates that for each pi bond that is added about 60 kcals of energy is added. The double bond has 146 kcal and when the 83 kcal for the sigma bond is subtracted 63 kcal are left. The triple bond has 200 kcal leaving 117 kcal/2 = 58.5 kcal per pi bond.

62. A The hybridized orbital is a combination of the s and p orbitals, therefore, the energy is between the energy of the combined orbitals. The s orbital has lower energy than the p orbital.

63. A The pi bond is formed by the overlap of two p orbitals.

64. B The sp^3 hybridized orbital has the combination of four orbitals, one s and three p orbitals, so one fourth (25%) is s character.

65. C Carbon 2 has four bonds so one s and three p orbitals combine to form the bonds leading to an sp^3 hybridization.

66. B Carbon 5 has three sigma bonds, so one s and two p orbitals combine leading to an sp^2 hybridization.

67. C Nitrogen 8 uses four orbitals (three for bonds and one for the lone pair), so one s and three p orbitals combine to form the bonds leading to an sp^3 hybridization.

68. C Carbons with sp^3 hybridization have bond angles of 109°.

69. B Carbons with sp^2 hybridization have bond angles of 120°.

70. C The sp^3 hybridized atom has bond angles of 109°.

71. C Carbon 9 has four bonds so one s and three p orbitals combine to form the bonds leading to a sp^3 hybridization.

72. C Nitrogen 8 uses four orbitals (three for bonds and one for the lone pair), so one s and three p orbitals combine leading to an sp^3 hybridization.

73. C Oxygen 3 uses four orbitals (two for bonds and two for lone pairs), so one s and three p orbitals leading to a sp^3 hybridization.

74. B Carbon 5 has three sigma bonds (two to carbon atoms and one to a hydrogen atom), so one s and two p orbitals combine leading to a sp^2 hybridization.

75. B Carbons with sp^2 hybridization have bond angles of 120°.

76. C The sp^3 hybridized atom has bond angles of 109°.

77. C Carbons with sp^3 hybridization have bond angles of 109°.

78. C Carbon 2 is sp^2 hybridized which means that one s and 2 p orbitals combine to form 3 sp^2 hybridized orbitals which means one third (33.3%) is s character.

79. D Nitrogen 8 is sp^3 hybridized so one s and three p orbitals combine to form four sp^3 hybridized orbitals which means three fourths (75%) is p character.

80. **C** The bond between an oxygen atom and carbon is polar covalent because of the electronegativity difference between the two atoms.

81. **C** The center carbon in acetone is sp^2 hybridized, so one s orbital and two p orbitals combine to form 3 sp^2 hybridized orbitals. This means that two thirds (66.6%) is p character. The structure of acetone is:

acetone

82. **D** To have delocalized electrons, there must be more than one double bond in an alternating single-double bond carbon chain. This is called *conjugation*.

83. **D** To have delocalized electrons, there must be more than one double bond in an alternating single-double bond carbon chain. This is called *conjugation*.

84. **D** The delocalized electrons of benzene stabilize the molecule by creating shorter stronger bonds between the carbons. The bonds are not as strong as double bonds, but they are stronger than single bonds.

85. **D** The delocalized electrons of benzene stabilize the molecule by creating shorter stronger bonds between the carbons. The bonds are not as strong as double bonds, but they are stronger than single bonds.

86. **D** Answer D would require moving a hydrogen so it is not a resonance structure.

87. **A** The dipole in ethanol results from the electronegativity difference between oxygen and carbon. Oxygen is more electronegative. The partial positive charge is on the carbon, so the dipole vector points toward the oxygen.

88. **D** Electronegativity differences between atoms create polar bonds. Carbon, chlorine, oxygen, and hydrogen all have electronegativity differences.

89. **B** Although structures I and II have polar bonds, they are oriented so that they cancel each other leading to a molecule that is not polar.

90. **B** The electronegativity difference is between the oxygen and the carbon, so the net dipole of the molecule is in the direction of the oxygen with the partial negative charge on the oxygen.

91. **D** Since the carbon-chloride bond has the largest dipole moment, it has the greatest electronegativity difference.

92. **A** The structure of carbon tetrabromide is tetrahedral and the bond dipole moments cancel to give a net dipole of zero.

93. **B** The electronegativity difference decides the distribution of electrons between bonded atoms. This distribution creates the polarity of the bond.

94. **A** Cyclohexane has no net dipole moment, so the only intermolecular interactions are London dispersion forces. Because the questions asks for the bonds between molecules covalent bonds (answer C) does not apply.

95. **D** Acetone has polar bonds so the primary interaction between the molecules is dipole-dipole. Because the questions asks for the bonds between molecules covalent bonds (answer C) does not apply. Hydrogen bonds may only occur when a hydrogen atom is bonded to fluorine, oxygen, or nitrogen. This is not the case in acetone.

96. **B** Diethylamine has a nitrogen-hydrogen bond, so it can hydrogen bond leading to hydrogen bonding as the primary interaction. Because the questions asks for the bonds between molecules covalent bonds (answer C) does not apply.

97. **B** A higher boiling point indicates that the interactions between the molecules are stronger. Hexanol has a hydrogen-oxygen bond, so hydrogen bonding is the primary interactions between the molecules. Hydrogen bonds are stronger than the London dispersion forces between hexane.

98. **B** Both isomers have the same molecular weights which eliminates answers C and D. If the *cis* isomer has the greater boiling point, it must have the greater dipole moment (answer B). The dipole moments support this observation because the dipole moments in the *trans* isomer cancel.

99. **A** The bonds between atoms are stronger than the bonds between molecules eliminating answers C and D. Sigma bonds are more stable than pi bonds, so they are harder to break.

100. **C** Triple bonds have three bonds to break so it requires the most energy.

101. **B** The hydrogen bond is the strongest intermolecular bond, so it requires the most energy to break.

102. **C** The question states that to be IR active the molecule must have a dipole moment. Dipole moments result when there are atoms with different electronegativities. Structures I and III have oxygen atoms which create a difference. Structure II is symmetric so there is no dipole moment.

103. **B** Only structure III is a conjugated system with alternating single and double bonds.

104. **D** Structure D would require moving hydrogens instead of just electrons.

105. **C** In order to be aromatic, a molecule must be planar, conjugated, and have 4n+2) pi electrons (Huckel's rule). Only answer C has a structure that has (4n+2) pi electrons. The lone pair of electrons count because they are in the p orbitals in the resonance system.

106. **A** The major resonance contributor structures minimize the formal charges on atoms. Answer A has the minimized formal charge.

107. C Answer C moves a hydrogen atom in addition to the electrons.
108. D Answer D does not meet the Huckel's rule: in order to be aromatic, a molecule must be planar, conjugated, and have (4n+2) pi electrons.
109. A Answer A has a structure that can resonance stabilize the negative charge onto both oxygen atoms. It is also the structure of a carboxylic acid.
110. A The resonance structure below shows the double bond on the nitrogen.

111. A The two structures have the same molecular weight ruling out answers C and D. If 2-nitrophenol is more volatile, the intermolecular interactions must be weaker making answer A the right answer.
112. B Only the double bonds contained in the ring system and the one double bond on the top of the ring structure are contained in the conjugated system. The bonds need to be alternating between single and double bonds. The conjugated system is circled on the Chlorophyll-a shown below.

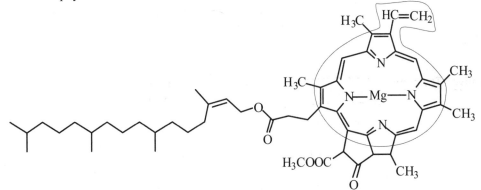

113. C Metal bonds with ligands in coordinate covalent bonds because the ligand donates a lone pair of electrons and the metal donates empty orbitals.
114. A There are no triple bonds in the system, so there are no sp hybridized carbons.
115. B The hybridization of the carbon can be determined from information in the question the trigonal planar structure and the bond angle both indicate sp^2 hybridization.
116. A The increased acidity indicates an increased stability. Answer B and D are in the opposite direction of the observation. Answer C is too general to be true. Leaving answer A as the correct answer. When a proton is removed from cyclopentadiene 6 electrons are now in the pi system.

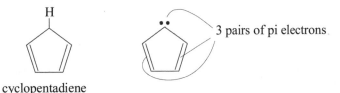

cyclopentadiene

117. B In the first structure the nitrogen requires a p orbital to form the double bond, leaving one s and two p orbitals to form 3 sp^2 hybridized orbitals. In the second structure, four sp^3 hybridized orbitals are formed with the one s and three p orbitals (one for the lone pair and 3 for the bonds).
118. B The length of the chain affects the fluidity because longer chains are able to form stronger London dispersion forces.
119. C Answer C is the only molecule that has a carbon atom with four different substituents on it.
120. A The only difference between the two structures is a rotation around a sigma bond indicating conformers.
121. C The two structures have the same molecular formula, but different connectivity.
122. D The two structures have the same connectivity, but different spatial orientation that are not mirror images. *Cis* and *trans* isomers are geometric isomers.
123. C The two structures have the same molecular formula, but different connectivity.
124. B The two molecules are non-superimposable mirror images of each other.
125. C The two structures have the same molecular formula, but different connectivity.
126. A The only difference between the two structures is a rotation around a sigma bond indicating conformers.
127. B This conformer has the two largest groups eclipsed.
128. C This conformer has the two largest groups in the anti position.

129. **B** The double bond cannot rotate, so there are no conformers for the double bond, however the two single bonds can rotate.
130. **B** The conformer has the two largest groups staggered, but located next to each other.
131. **B** The conformer has the two largest groups staggered, but located next to each other.
132. **D** The conformer has the atoms eclipsed, but the largest groups are not eclipsing each other (fully eclipsed).
133. **D** The two structures have the same connectivity, but different spatial orientation. The two structures are *cis* and *trans* isomers.
134. **D** Structures I and III are chiral molecules and therefore, they are optically active. Structure II has two Br atoms connected to the central carbon.
135. **A** The Br atom has the highest atomic weight and the therefore, the highest priority.
136. **B** To determine priority, one must determine which atom connected to the central carbon has the highest atomic number. If there is a tie, one must look at the next group of connectivities. In this case, there are three carbon atoms connected to the central carbon. One of these carbons is connected to an oxygen that has the highest atomic number; thus, this group is the highest priority. The ethyl group is a higher priority than the methyl group because its first carbon is attached to another carbon which has higher atomic number than the three hydrogens attached to the methyl carbon.
137. **B** See the answer to question 136.
138. **A** The two structures have the same connectivity, but different spatial orientation. They are not mirror images, but they are *cis* and *trans* isomers.
139. **D** The two molecules do not have the same connectivity, so they are not stereoisomers. They are structural isomers.
140. **C** D-Glucose and D-Mannose differ only at carbon number 2 making an epimer. (Epimers are diastereomers that differ at only one chiral carbon.) Epimer is a better description than diastereomer.
141. **A** D-Glucose and D-Altrose are stereoisomers that are not mirror images and they are different at more than one carbon.
142. **A** D-Mannose and D-Galactose are stereoisomers that are not mirror images and they are different at more than one carbon.
143. **C** D-Altrose and D-Mannose differ only at carbon number 3 making an epimer. Epimer is a better description than diastereomers.
144. **C** Each intersection in D-Glucose is a chiral center giving 4 chiral centers.
145. **D** The maximum number of stereoisomers is 2^n where n is the number of chiral centers which equals $2^4=16$.
146. **B** The molecule in the question is the mirror image of D-Glucose, therefore it is an enantiomer.
147. **C** All are diastereomers because none of them are mirror images and all are stereoisomers. Epimers are diastereomers.
148. **A** There are no chiral centers because there are no sp^3 carbons with four different substituents.
149. **A** A *trans* double bond is converted to a *cis* isomer.
150. **D** All of the double bonds are conjugated including the carbonyl.
151. **D** The chiral centers are circled on the structure below.

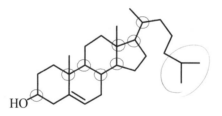

152. **D** The maximum number of stereoisomers is equal to $2^8 = 256$. Answer D is the only one greater than 2^3.
153. **C** The higher priority groups are on the same side of the double bond, therefore, it is Z.

154. B The priorities are shown below. The priorities go clockwise, but the absolute configuration is S because H, the substituent with the lowest priority, is coming out of the page, and absolute configuration is determined when the lowest priority group is going into the page.

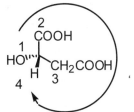

155. B The priorities are shown below. The priorities go clockwise, but the absolute configuration is S because H is coming out of the page.

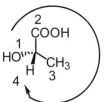

156. C The direction that a chiral center rotates light cannot be determined unless it is measured.

157. A Both chiral centers have the priorities rotating counterclockwise (below), but the hydrogens are coming out of the page, so both have an absolute configuration of R.

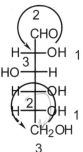

158. A The priorities rotate counterclockwise (below), but the hydrogen is coming out of the page.

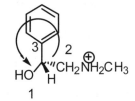

159. B The (l) in the name indicates that the molecule rotates light counterclockwise.

160. B The molecule is a mirror image of (l)-epinephrine with the absolute configuration of S.

161. A The enantiomer rotates light in the opposite direction of the original molecule.

162. C The first double bond on top of (+)-carvone is Z because the two higher priority substituents are on the same side of the double bond. The double bond that is on the side chain has two hydrogens on one end, so there are no geometric isomers.

163. B The two molecules are mirror images of each other. It can also be determined from the fact that (+) and (-) molecules are enantiomers.

164. C Because the molecules are enantiomers answers A and D are ruled out. The groups of (+)-carvone are oriented in a counter clockwise direction so its absolute configuration is S, therefore, (-)-carvone is R.

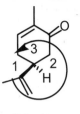

(+)-carvone

165. A The (+) in the name indicates that the molecule rotates light clockwise.
166. B The only physical property that is different for enantiomers is the direction of rotation of plane polarized light.
167. A (d) is the same as (+). Both rotate plane polarized light clockwise.
168. B Both the higher priority groups, the Br atoms, are on the same side of the ring.
169. C The priorities are counter-clockwise on carbon 1 and clockwise on carbon 2.

170. D The product is a meso compound because it has a plane of symmetry. It will not rotate polarized light.
171. C There is a plane of symmetry down the middle of the molecule.

172. A The two structures are stereoisomers that are not mirror images.
173. B The two highest priority groups are on the opposite sides of the ring so it is *trans*.
174. C There are two reasons why solutions of chiral molecules don't rotate light (I and II).
175. B Any process that separates enantiomers is called resolution.
176. A There are no sp^3 carbons in allene, so there are no chiral centers.
177. B The two structures are non-superimposable mirror images.
178. D These two structures are the same molecule viewed from two angles separated by 90°.
179. B Enantiomers rotate light the same magnitude in the opposite direction.
180. B There is more (-) than (+) so the net result is counterclockwise rotation.
181. A Both carbons have the priorities oriented in the counterclockwise direction, but both hydrogens are coming out of the page, so their absolute configurations must be R.
182. D Since (-)-tartaric is the enantiomer of (+)-tartaric the absolute configurations of all stereogenic centers are reversed.
183. B The (+) and (-) indicates enantiomers.
184. C The enantiomer rotates light the same magnitude in the opposite direction.
185. A The structures are stereoisomers that are not mirror images.
186. B There is a plane of symmetry through the center that makes the compound meso and thus optically inactive.
187. B Both (-) and (l) indicate that the compound rotate, light counterclockwise.
188. B Two isomers (+) and (-) are optically active, but the third isomer is a meso compound.
189. D The only physical property that can be different between enantiomers is the direction in which they rotate plane polarized light.
190. B Answers A and D are ruled out because they still contain Br, and the question states that Br is replaced. Since the Cl is on the opposite face, the product is (R)-chlorobutane, which is answer B.
191. D A racemic mixture is a 50:50 mixture of enantiomers which does not rotate light.
192. A A racemic mixture is a 50:50 mixture of enantiomers which does not rotate light.
193. A The change from (l)-dopa is the loss of the COOH groups. No amide is present in either product ruling out answers B and C.
194. B The group priorities are oriented counterclockwise giving an S absolute configuration.
195. B Both (l) and (-) indicate a molecule that rotates polarized light counterclockwise.
196. B The (l) indicates that it rotates light counterclockwise.
197. A There are no sp^3 carbons with four different substituents.

198. D Dopamine is not optically active because it has no chiral centers.

199. C Answer C is the only one that is (d)-dopa the rest are (l)-dopa rotated.

200. B Enantiomers cannot be separated by distillation (same boiling point) and recrystallization, so answer III is the only one that works.

201. D There are three chiral centers starred below.

202. D The maximum number of optically active isomers is 2^n, where n is the number of chiral centers, so 2^3 is 8.

203. C The change removes one chiral center by shifting a double bond.

204. A Answer A is the same molecule not an isomer.

205. C The statement in the question makes it clear that 1-phenylethanol must be chiral because it is optically active. Draw the structures and test it.

206. D Answer A and B are ruled out because the two are enantiomers. (+)-1-phenylethanol has the priorities oriented counterclockwise as written, but the hydrogen is coming out causing (R). (-)-1-phenylehthanol must be the opposite.

207. A The fact that there is some (+) rotation indicates that there must be more (+) isomer present. However, there is still a mixture because the rotation is not +42, the amount of rotation of the pure enantiomer.

208. A Answer A has S absolute configuration and matches the name in the question.

209. C Ethylacetoacetate has no chirality ruling out answers A and B. The (+) indicates the clockwise direction.

210. A   Alkanes contain only sp^4 hybridized carbons.

211. B   Answer B is an alkene because it contains a double bond.

212. D   Answer D contains an O atom, so it must have a functional group. Although some of the other answers are deficient in hydrogens, they could be cycloalkanes.

213. A   Haloalkanes are alkanes with halogens and answer A is the only answer with halogens.

214. B   Geminal halogens have two halogens on the same carbons.

215. B   C1 has only one other carbon attached to it.

216. D   C4 has three carbons attached to it.

217. C   C3 has two carbons attached to it.

218. C   The tertiary carbons a circled on the structure below.

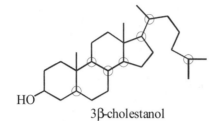

3β-cholestanol

219. B   The primary carbons are circled below.

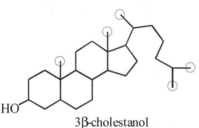

3β-cholestanol

220. D   Answer D is a structure with a single electron (radical).

221. A   A tertiary radical is the most stable radical. A more substituted radical carbon is more stable.

222. D   Answer D is a primary radical so it is the least stable. A more substituted radical carbon is more stable.

223. C   The carbon labeled 3 is the most reactive because it forms the most substituted radical. The carbon labeled 1 does not have a hydrogen to abstract.

224. D   Answer D has the lowest numbers on the substituents and names the longest chain.

225. A   Answer A has the lowest numbers on the substituents and the substituents numbers start at the point of connection (i.e., 3 not 2 ethyl).

226. A   For straight chain alkanes, the longer the chain (the higher the molecular weight), the higher the boiling point. The reason that the longer chain has a higher boiling point is that there is more surface area available for London dispersion forces.

227. C   With alkanes that have the same number of carbons, increased branching decreases the boiling point. Answer C has the most branching.

228. C   The smaller the molecular weight, the lower the density.

229. C   For straight chain alkanes, the smaller the molecular weight (chain length), the lower the melting point because the surface is smaller leading to fewer London dispersion forces.

230. A   Although combustion reactions need energy to get started, they produce energy (heat) once they are started.

231. D   All three are steps of a radical halogenation.

232. C   Alkanes are less reactive than ethers, aldehydes, and anhydrides.

233. B   Octane is the only alkane with a longer chain (i.e., the higher molecular weight), and thus a higher boiling point.

234. A   Dodecane is the only alkane with a longer chain (i.e., the higher molecular weight) which creates a higher boiling point. Branching lowers the boiling point.

235. C   When alkanes are reacted with oxygen at high temperatures, combustion takes place producing CO_2 and H_2O.

236. B   Oxygen is required because it is a reactant. The high temperatures are required to provide the activation energy.

237. A   Pentane is the shortest chain (i.e., lowest molecular weight), so it has the lowest boiling point.

238. B   Octane is a longer carbon chain, and has a higher boiling point, so it must be liquid or gas. You can remember that octane a major component of gasoline for automobiles, and is a liquid at room temperature.

239. C   The density must be between pentane and octane and answer C is the only one that is in between the other two densities.

240. A Propane has a lower boiling point than butane, so propane would have to be a gas at room temperature.
241. A Hydrogen is not a product of combustion.
242. D Based on the balanced equation below, the combustion of butane produces 5 moles of H_2O.

$$C_4H_{10} + \text{}^{13}/_2O_2 \rightarrow 4CO_2 + 5H_2O$$

243. C Based on the balanced equation below, the combustion of octane produces 8 moles of H_2O.

$$C_8H_{16} + 12O_2 \rightarrow 8CO_2 + 8H_2O$$

244. C Based on the balanced equation below, the combustion of heptane produces 7 moles of CO_2.

$$C_7H_{16} + 11O_2 \rightarrow 7CO_2 + 8H_2O$$

245. D Based on the balanced equation below, the combustion of cyclopropane produces 3 moles of CO_2.

$$C_3H_6 + \text{}^9/_2O_2 \rightarrow 3CO_2 + 3H_2O$$

246. C Based on the balanced equation below, the combustion of nonane consumes 14 moles of O_2.

$$C_9H_{20} + 14O_2 \rightarrow 9CO_2 + 10H_2O$$

247. B Based on the balanced equation below, the combustion of cyclohexane consumes 9 moles of O_2.

$$C_6H_{12} + 9O_2 \rightarrow 6CO_2 + 6H_2O$$

248. C Octane has the most carbon and hydrogen atoms so the reaction will produce the most heat.
249. B Octane will have a melting point between hexane and decane and answer B is the only one between the two values given in the question.
250. B Isopentane will have a boiling point a little smaller than pentane because branching reduces the boiling point of an alkane.
251. C Cyclohexane will have a boiling point between cyclopentane and cyclooctane.
252. B Choice B arranges the alkanes in the order of increasing molecular weight, which corresponds to the increasing densities.
253. B The iso alkanes are branched, so they have a lower boiling point than the straight chain alkanes.
254. D The chair conformation is shown.
255. A The boat conformation is shown.
256. A The chair conformation is the most stable.
257. A Six member rings have the least strain and five member rings are the second most stable rings.
258. D Alkanes only undergo two of these types of reactions, combustion and radical.
259. C Like dissolves like, so a long chain alkane can only be dissolved in a nonpolar solvent. Heptane is the only solvent in the list that will work.
260. C Neither alkane has hydrogen bonding so answers A and B are eliminated. According to the boiling points in the question decane must have the stronger forces because high boiling points indicate strong intermolecular forces.
261. A All of the structures in the table have the same molecular weight, so answers C and D are eliminated. Since the boiling points are decreasing with increased branching, the London dispersion forces must be decreasing.
262. C Both molecules do not hydrogen bond because there is no H-F, H-N, or H-O bonds, eliminating answer A. Since both molecules are about the same size, answers B and D are eliminated. The 1-fluoropropane has a dipole because of the large difference in electronegativity between fluorine and carbon leading to stronger intermolecular forces and a higher boiling point.
263. D The larger the alkane chain the stronger the London dispersion forces leading to a solid. Answer D is the longest alkane chain in the answers. The alkanes in answers A, B, and C are all liquids.
264. C Answer C is the only answer that would be less likely to form a solid. Answers A and B hydrogen bond and mineral oil is longer than diesel fuel. The idea is that an additive with a lower melting point needs to be used.
265. C Flaring is a combustion process so the products would be water and carbon dioxide.
266. A In an initiation step, there are no radicals in the reactants and a radical is formed in the products. Answers B and C are propagation steps and answer D is a termination step.
267. D In a termination step two radicals combine to form a product that is not a radical. Answers B and C are propagation steps and answer A is an initiation step.
268. B Homolytic cleavage is the formation of two radicals from a single bond. Only the initiation step has homolytic cleavage.
269. D Answer D is a termination step because a radical is not present in the products.
270. C Termination is the only step that does not have a radical in the product.
271. D The ultraviolet light is only required to start the reaction, then propagation keeps the reaction going.

272. **C** The bond labeled 3 requires the least amount of energy because it forms the most stable radical product, a tertiary radical.

273. **A** Answer A gives the correct order of reactivity. Fluorine is the most reactive radical and least selective.

274. **B** Even though the tertiary hydrogen can react at a faster rate, there are nine times as many primary hydrogens with which the chlorine can react. The product ratio indicates that the tertiary reaction rate is not nine times faster than the primary.

275. **A** H_2O_2 is the best choice because the bond dissociation energy of HO-OH is the smallest in the table.

276. **D** Answer D is the only answer that makes sense. Both butyl and allyl radicals are relatively stable radicals, eliminating answers A and B. Sterics is not a concern for radical reactions.

277. **A** The hydrogen shown is axial because it is not in the plane of the ring.

278. **B** The hydrogen shown is equatorial because it is in the plane of the ring.

279. **C** The two substituents are on the same side of the ring. If the hydrogens are drawn in, they both sit below the two methyl groups shown (for 1,2 substitution on a cyclohexane ring axial, equatorial is a cis relationship).

280. **D** The most stable configuration occurs when the largest groups are in the equatorial position. Answer C is a 1,2 substitution and answer B is trans.

281. **A** Answer A has the larger group in the equatorial position. Answer B and C are trans configurations.

282. **C** Answer C is the trans configuration that has the larger groups in the equatorial position.

283. **D** The table shows that the rate of reaction corresponds with reactivity of the halogens. Fluorine is the most reactive radical and it has the fastest rate of reaction.

284. **C** All four products are shown below.

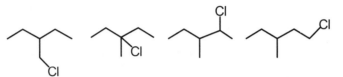

285. **B** Bromine is more selective so only one product results. The product is shown below.

286. **B** The question states that the product is monobrominated so answers C and D are eliminated. Answer B has the product that involves the most stable radical.

287. **D** The chlorine radical is less selective, so all three products are possible.

288. **B** Radical reactions are hard to control because they propagate and form more than mono substituted products.

289. **A** Fluorine radicals are so reactive that they are not selective. They will react most often with whichever hydrogen type is the most prevalent. There are six of this type of hydrogen available to be substituted in answer A.

290. **C** The bonds need linear overlap to be stable so III is eliminated. I and II are both true.

291. **C** The heat of combustion coordinates with ring strain in the table. Cyclohexane has the lowest ring strain and the lowest heat of combustion.

292. **D** The fuel is the alkane and the air is the source of oxygen which are both reactants. Water and carbon dioxide are both products.

293. **D** Alkenes contain double bonds.

294. **C** Alkynes contain triple bonds.

295. **D** The stability of an alkene increases with substitution.

296. **A** The stability of an alkene decreases as the degree of substitution decreases.

297. **C** The longest chain is seven carbons, a heptene. Answer C gives the lowest numbered carbon attached to the double bond.

298. **A** The longest chain is six carbons, and the triple bond gets the lowest number carbon. The suffix -yne indicates a triple bond.

299. **A** The substitution pattern is ortho because the two substituents are adjacent to each other..

300. **A** The substitution is 4 so it must be on the opposite side of the benzene ring giving a para substitution. Even if you don't know ansiate, you should know that it must be para (p).

301. **C** The first bond has the higher priority groups on opposite sides of the double bond (E). The second bond has the higher priority groups on the same side of the double bond (Cl and the chain).

302. **B** The bond between C1 and C2 has two hydrogens on one end so it cannot have geometric isomers. The second double bond has the Cl and methyl group (highest priority) on the same side of the double bond, which is Z assignment.

303. **C** Both alkanes and alkenes have sigma bonds so answers A and B are eliminated. The pi bond in the alkenes must stabilize a negative charge if they are more acidic, answer C.

304. **C** The boiling point must be between 1-butene and 1-hexene, and answer C is the only one in between the two.

305. B The para substitution (1,4) has two axes of symmetry, while ortho and meta substitutions only have one. This extra symmetry will cause it to have a higher melting point.

306. C 1-Butene and 1-butyne have very similar molecular weights, eliminating answers A and B. If 1-butyne has the higher boiling point, then it must have the stronger intermolecular forces which results from a more polar molecule.

307. A Given the information in the questions 1-butyne is a gas at room temperature and 1-propyne would have an even lower boiling point, so it must also be a gas.

308. A Cations follow the same pattern as radicals. Increased substitution leads to increased stability of the cation.

309. B Cations follow the same pattern as radicals. Increased substitution leads to increased stability of the cation.

310. B Halogens are electron withdrawing. Memorize which groups are electron withdrawing and which are electron donating.

311. D Methoxy groups are electron donating. Memorize which groups are electron withdrawing and which are electron donating.

312. B 3-Hexyne is a triple bond so hydrogenation with one mole of hydrogen leads to one mole of 3-hexene. The alkene will be cis because the hydrogen adds in a syn fashion.

313. A If 3-hexyne is hydrogenated with excess hydrogen, the alkane is formed.

314. A A metal catalyst is required for hydrogenation because of the high activation energy.

315. C Both Ni and Pt are metal catalysts used for hydrogenation.

316. C Two pi bonds using one mole of H_2 each, so 2 moles of hydrogen will be consumed in the conversion.

317. D According to the table, 2,3-dimethyl-2-butene is the most stable alkene because it has the lowest heat of hydrogenation. This follows the rule that the most substituted alkene is the most stable.

318. A According to the table, ethylene is the least stable because it has the highest heat of hydrogenation. This follows the rule that the most substituted alkene is the most stable.

319. B The trend observed for 2-butene would hold for 2-pentene. Since the heat of hydration for *cis*-2-pentene is the same as *cis*-2-butene, the hydration of *trans*-2-pentene should be the same as that of trans-2-butene.

320. C As the substitution on the alkene carbon is increased, the heat of hydrogenation decreases.

321. B The lower heat of hydrogenation for the *trans* isomer indicates that it is more stable ruling out answers A, and C. Molecular weight has nothing to do with bond stability and both isomers have the same molecular weight ruling out answer D. Answer B makes sense. When the larger substituents are on the same side of the double bond there is steric interaction which decreases the stability of the molecule.

322. D Answer B is true, but it does not explain the difference. Answer A is wrong and answer C is counter to what the question states. Answer D explains the difference logically.

323. B The double bonds in benzene are less reactive than in a regular alkene because hydrogenation ruins the aromaticity of the ring. Answers A, C, and D are all wrong.

324. C The addition of 4 hydrogens indicates that there were two double bonds.

325. C The addition of three moles of hydrogen adds six hydrogen atoms to the formula.

326. A The addition of four hydrogen atoms indicates two double bonds, so answer A is the only correct answer. The fact that the saturated alkane is still 4 hydrogens short (would be 32 hydrogens if there were no rings) indicates two rings in the structure.

327. A A captial E indicates that the reaction mechanism is an elimination.

328. B The 1 indicates that the rate law is first order. There is only one molecule involved in the rate determining step.

329. A An E1 reaction forms a double bond by breaking two single bonds.

330. C In an electrophilic addition the double bond is converted to two single bonds.

331. C An eletrophile "loves electrons", so the only species listed that will accept electrons is the H^+ ion.

332. B Benzene does not undergo addition reactions because they would result in one double bond being converted to two single bonds, thus disrupting the aromaticity. With substitution the aromaticity is conserved.

333. C The mechanism for the reaction is shown below.

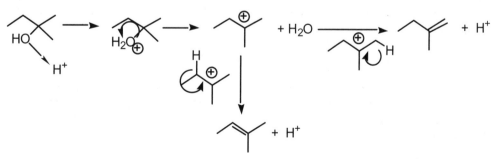

334. B The proton from the sulfuric acid is a catalyst. See mechanism above.

335. A The bromine will add to the more substituted side of the double bond because the cation intermediate that is formed will have the positive charge and the most substituted carbon of the double bond. This is Markovnikov's rule. See the mechanism below:

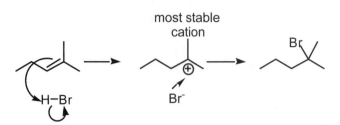

most stable cation

336. D As shown in the mechanism in the answer to question 333 a carbocation is formed in the first step of the mechanism. There is free rotation around the sigma bonds allowing all three products to be formed.

337. C Ozone (O_3) oxidizes the alkene by adding oxygen to both ends of a cleaved double bond. It forms aldehydes or ketones. This reaction is called ozonolysis.

338. B An alkene and alkane are lower oxidation states than an alkyne, so hydrogenation reduces an alkyne.

339. D Dilute cold acid will promote hydration. The water in the acid is the source of the hydroxyl group.

340. D The 2-methyl-2-butene will react the fastest because it forms the most stable carbocation intermediate.

341. C The benzene ring in answer C has an electron donating group which activates the ring to electrophilic substitution.

342. B The benzene ring in answer B has an electron withdrawing group which deactivates the ring to electrophilic substitution.

343. B Ozonolysis always forms an aldehyde or a ketone.

344. A The double bond is protonated to form the most stable carbocation which is shown below. The water then attacks the carbocation to form the alcohol.

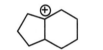

345. C The Br group on the benzene ring is ortho-para directing and slightly deactivating.

346. B The carboxylic acid group is meta directing and deactivating.

347. C The hydroxyl group is ortho, para directing and activating.

348. B Following the mechanism shown in the answer to question 335, the most stable carbocation is formed.

349. A The $FeBr_3$ activates the Br_2 so that the electrons in the benzene ring can attack the electrophile.

350. A The bond labeled 1 is the most substituted double bond, so it is the most stable.

351. B Three double bonds require three moles of hydrogen.

352. A All of the double bonds have two substituents on one end of the bond that are the same. Consequently, there is no possibility for geometric isomerism.

353. D The breaks in the double bonds are shown below. In ozonolysis, an oxygen is added to each side of the broken double bond. Note that there is one product with three oxygens, two of which are on neighboring carbons. No product has four oxygens.

354. B The first step of an E1 mechanism involves the formation of a carbocation after protonation of the alcohol. The alcohol in answer B will form the most stable carbocation.

355. A Dehydration is promoted by concentrated hot acid.

356. D Hydration is promoted by dilute acid under cold conditions.

357. B Hydrogenation is achieved using hydrogen and a metal catalyst.

358. A Bromine adds in an anti conformation because the first bromine atom forms a bromium ion on one side of the ring. The second bromine atom attacks the bromonium from the backside, creating an anti conformation.

359. B Four moles of hydrogen would be required to form a saturated alkane. Each double bond requires one mole of hydrogen in order to be saturated.

360. C Acetone would result from the ozonolysis of bond number 1.

361. D The backbone of each of the products is circled below.

362. B Answer B has the Br atoms on the most substituted carbon of each double bond.

363. D The bond labeled 4 forms the least stable cation interdediate in the reaction with HBr, causing it to be the slowest to react.

364. C The structure in answer C has the hydroxyl groups on the most substituted carbons of the double bonds because the water attacks the carbocation that forms. The most stable carbocation is the one on the most subsituted carbon.

365. C The reaction involves an electrophilic attack on a double bond.

366. A The carbocation formed in the first step of the reaction allows for several alkene products due to hydride shifts and methyl shifts.

367. C The hydrogenation of the double bond adds hydrogen in a syn fashion which forces the two substituents to be on the same face of the ring.

368. B The hydrogen can add to either face of the double bond so the substituents can be on the same face as the third methyl group or on the opposite face.

369. D As with dehydration, dehyrohalogenation begins with the formation of a carbocation which allows the formation of several products depending on which hydrogen is removed. Since the intermediate carbocation is planar, both *cis* and *trans* isomers may be formed.

370. A The question states that the addition creates the anti-Markovnikov product; the Br adds to the less substituted carbon.

371. A The fact that the addition is anti-Markovnikov indicates that a carbocation is not formed. Radical reaction is the only other mechanism that makes sense. Peroxide is another clue that the reaction involves a radical.

372. A The fact that the addition is anti-Markovnikov indicates that the hydroxyl group will add to the least substituted carbon. Choices B and C are named incorrectly.

373. C The fact that the reaction mechanism begins with the addition of an electrophile to a double bond indicates that it must be an electrophilic addition.

374. C The addition of a halogen is in the anti fashion because of the cyclic chloronium ion that is formed in the first step. This is similar to the bromonium ion of question 358.

375. C The addition of a halogen is in the anti fashion because the cyclic bromonium that is formed. See question 358.

376. B   The formation of the carbocation allows for the formation of all of the products except answer B.

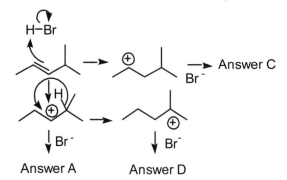

377. C   The formation of the carbocation leads to attack by the chlorine atoms from either side.

378. C   The 2-pentenes are more substituted therefore more thermodynamically stable than 1-pentene.

379. B   The cis isomer has steric hindrance due to the larger groups being on the same side of the double bond, so this isomer is less stable.

380. C   The products require the removal of a hydrogen and a bromine atom (dehydrohalogenation).

381. B   The 1,4-addition product is more stable because it is more substituted than the 1,2-addition product.

382. B   The carbocation that is formed allows resonance stabilization.

383. A   The 1,2-addition product is kinetically favored so at kinetic conditions (low temperatures), it is the favored product.

384. B   The bromine adds in an anti fashion, but there is no resonance of the cation. There will be two stereoisomers.

385. C   There are 2 structural isomers because the allyl cation has two resonance structures shown below leading to two different products.

386. A   There are zero moles of acetone produced for each mole of 1,3-butadiene. Each end of the molecule produces a mole of formaldehyde, not acetone.

387. A   The ketyl group substitutes for a hydrogen on the aromatic ring. It is an aromatic substitution reaction.

388. C   The methoxy group is ring activating. The other substituents are deactivating. Memorize the table of activating and deactivating groups.

389. D   The symmetry of the ring causes the substitution to produce only one product.

390. C   The deactivation of the ring by the ketyl group prevents the second substitution of the ring.

391. A   The product is meta, so the substitution is meta directed eliminating answers B and C. Meta directing groups on the whole are deactivating. Memorize the chart.

392. B   The amine group is activating.

393. C   The hydroxyl group is ortho-para directing producing two products.

394. D   The definition of a substitution reaction is the exchange of one sigma bond for a new sigma bond.

395. D   The definition of a substitution reaction is the exchange of one sigma bond for a new sigma bond.

396. B   The capital letter S stands for substitution.

397. C   A S_N1 reaction forms a carbocation in the first step of the mechanism. In the second step, a nucleophile attacks the carbocation.

398. B   The only step of the S_N2 reaction is a substitution occurring with the new bond and old bond forming and breaking simultaneously.

399. C   The concentrations of both the nucleophile and the electrophile control the rate of the reaction because the rate determining step is bimolecular.

400. B   Only the concentration of the electrophile controls the rate of the reaction because the rate determining step is the unimolecular formation of the carbocation.

401. B   Step I is the rate determining step, so it must be slower than step II.

402. C   A nucleophile is an electron rich substance, thus a molecule with a negative charge is generally a good nucleophile.

403. A   A nucleophile is an electron rich substance, thus a molecule with a negative charge is generally a good nucleophile.

404. **A** Stable molecules are the best leaving groups. If all the leaving groups are charged, the iodide ion is the best leaving group. Memorize the order of the halogens for leaving groups and nucleophiles.

405. **A** Water is a good leaving group.

406. **B** A nucleophile donates electrons, so it must be a Lewis base.

407. **D** The reaction is an S_N1. Thus, whichever substrate can form the carbocation the fastest will react the fastest. Iodide is the best leaving group, forming the cation the fastest.

408. **A** The reactants are a primary halogen and a strong nucleophile leading to S_N2. S_N2 reactions are preferred when there is no steric hindrance.

409. **D** The reactants are a tertiary halgoen and a weak nucleophile leading to S_N1.

410. **A** Alcohols follow the same trend has alkanes. Ethanol is the shortest chain (i.e., lowest molecular weight) leading to the lowest boiling point.

411. **D** Alcohols follow the same trend has alkanes. Pentanol is the longest chain (i.e., highest molecular weight) leading to the highest boiling point.

412. **B** 1-butanol will have a boiling point in the middle of 1-propanol and 1-pentanol. Answer B is right in the middle. Answer A is too close to 1-propanol.

413. **D** 1-hexanol will have a higher boiling point than 1-pentanol and answer D is the only answer that is higher. The other boiling point given is irrelevant.

414. **A** Alcohols of the same chain length as alkanes, alkenes, and alkynes have higher boiling points due to their ability to hydrogen bond.

415. **B** Alcohols have a higher boiling point than alkanes so the alcohol would not have boiled and would still be a liquid.

416. **D** Ethers follow the same trend as alkanes, so diisopropyl ether will have the highest boiling point because it has the highest molecular weight.

417. **D** Alcohols have higher boiling point due to hydrogen bonding.

418. **C** The hydrogen bonding in cyclohexanol raises its melting point causing it to be a solid at room temperature.

419. **D** Although an ether cannot hydrogen bond, they do have a dipole moment causing the intermolecular forces to be stronger for ethers, which raises their boiling point.

420. **A** Alcohols can hydrogen bond with water, increasing their solubility.

421. **A** The oxygen in diethyl ether can hydrogen bond with water increasing the solubility of the ether.

422. **A** Short chain alcohols are very soluble in water due to hydrogen bonding.

423. **A** The shorter the chain of the alcohol, the more soluble it is in water (remember that like dissolves like).

424. **D** The higher electronegativity of the oxygen atom in an alcohol helps to stabilize the negative charge of the conjugate base of the alcohol. A more stable conjugate base means a stronger acid.

425. **C** The structure is shown below. There are three carbons attached to the carbon with the hydroxyl group.

426. **A** Phenol has resonance stabilization of the negative charge making it more acidic than the other alcohols.

427. **D** Answer A is not polar and answer B and C have acidic hydrogens. Only answer D is both polar and does not contain an acidic proton making it a good choice for a Grignard solvent.

428. **A** The ethanol molecule's oxygen acts as a nucleophile attacking the sulfur atom while displacing the chloride ion.

429. **A** A nucleophile forms a new sigma bond replacing the old sigma bond.

430. **C** A ketone is a higher oxidation state than an alcohol. Potassium chromate is an oxidant. Although ozone produces an oxidized product, it only works on alkenes.

431. **B** An aldehyde is a higher oxidation state than an alcohol. The oxidation has to be done carefully so that the aldehyde is not oxidized all the way to a carboxylic acid.

432. **C** The carbon atom of a carboxylic acid is at a higher oxidation state than the carbon atom in either an alcohol or an aldehyde. Potassium chromate is a strong oxidant. Although ozone produces an oxidized product, it only works on alkenes.

433. **C** The reaction is a S_N1 reaction producing a carbocation that can be attacked from either face producing a racemic mixture. The substitution reaction produces a new chiral center ruling out answers A and D.

434. **A** Since the Lucas reaction is a S_N1 reaction with a tertiary alcohol, the rate is only dependent on the concentration of the electrophile (the t-butyl alcohol carbocation).

435. **A** According to the table, primary alcohols are the slowest to react, and 1-pentanol is the only primary alcohol.

436. **D** Both secondary and tertiary alcohols react by a S_N1 mechanism, so the faster rate of reaction of tertiary alcohols results from the more stable cation.

437. **A** The $ZnCl_2$ is a Lewis acid catalyst. A Lewis acid would destabilize a nucleophile such as the chloride ion from HCl, therefore answer C is eliminated. Answer D is wrong and answer B does not make sense because HCl provides a better source of free chloride ion.

438. D The reactant is a tertiary alcohol, so the reaction will be a S_N1 mechanism which indicates the formation of a carbocation. Answer C is incorrect because the chart shows that tertiary alcohols react faster than primary alcohols. Answer A is incorrect because the product would be a racemic mixture. See question 433.

439. A The reaction is a S_N2 reaction that produces answer A.

440. A According to the table, only tertiary alcohols react that quickly.

441. C Since the alcohol is a tertiary alcohol, the rate only depends on the electrophile (S_N1).

442. A According to the table, only primary alcohols take over 6 minutes to react, so the alcohol must be 1-butanol since it is the only primary alcohol listed.

443. A PBr_3 is used to produce bromoalkanes from alcohols.

444. A The reaction described requires a proton to protonate the ether and form a leaving group. Only answer A is an acid.

445. A 3-hexanol is a lower oxidation state than 3-hexanone and only answer A is a reducing agent.

446. C Hydroxide is not a good leaving group, however, water is a good leaving group. The acid protonates the alcohol and forms water.

447. A $LiAlH_4$ is a reducing reagent that reduces aldehydes to primary alcohols, and ketones to secondary alcohols.

448. A The product is an aldehyde, which is more oxidized than an alcohol. PCC must be an oxidizing agent.

449. C To convert the aldehyde back to an alcohol a reducing agent must be used. $LiAlH_4$ (answer C) is the only reducing agent listed.

450. A An acid catalyst is required to convert hydroxide to a good leaving group. Answer A is the only one with an acid catalyst.

451. B Triphenylmethanol would react faster because the carbocation produced in the first step of the reaction would be stabilized by the phenyl rings.

452. B The reaction has to be S_N1 because the electrophile (triphenylmethanol) is too sterically hindered for S_N2. Answers C and D are eliminated because the reaction is a subsitution.

453. A The sulfuric acid converts triphenylmethanol to a stable carbocation. Once dissolved in methanol, the methanol attacks the carbocation to form answer A.

454. B The methanol oxygen attacks the carbocation, then a proton is lost to form the ether product.

455. A The structure shown in the question is an alcohol that was reduced from the original ketone. Only answer A is a reducing agent.

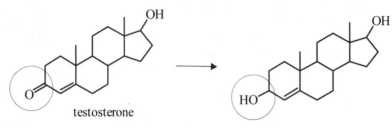

testosterone

456. C The structure shown in the question is a ketone that was oxidized from an alcohol. Answer C is the proper oxidizing agent. Ozone does produce an oxidized product, but it acts on alkenes.

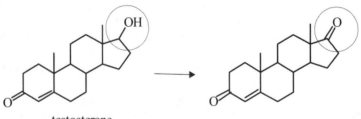

testosterone

457. B HBr is used to do an S_N1 substitution of an alcohol with a bromide ion. Therefore, a mixture of products would result.

458. B The reaction is a substitution reaction with the same nucleophile. The only difference between the two reactions is the leaving group. Since the question states that precipitation happens with 1-bromobutane faster, the bromide ion must be a better leaving group than the chloride ion.

459. A This is a primary haloalakane with a good nucleophile, so the reaction will be S_N2.

460. C Since 1-bromobutane proceeds by a S_N2 mechanism the rate depends on the concentration of the nucleophile. However the 2-chloro-2methylpropane proceeds by S_N1 so the rate only depends on the concentration of the electrophile and not the nucleophile (I^-).

461. C I^- is a better leaving group than OH^- so the substitution will be for I^-, eliminating answer B and D. The mechanism is S_N2 so the absolute configuration of the product must be inverted. Therefore, the new configuration is R (answer C).

462. A The answer must contain a syn (same side) diol. This eliminates choices B and D. Answer A is correct because it contains the two new alcohol groups produced in the reaction plus a third alcohol group that was present in the starting material.

463. D Answer D is the only product with the alkene cleaved.

464. A The question states that the product is optically active, so the mechanism must be S_N2 because S_N1 produces a racemic mixture of products (due to the carbocation intermediate). The reaction does not involve the formation of a double bond, so it is not an elimination (answers C and D).

465. C The reaction forms a double bond so answers A and B are eliminated. The fact that two alkenes were produced indicates that a carbocation is formed, so the reaction must be E1.

466. D The question states that the reaction is an oxidation. Only answer D, an aldehyde, is more oxidized than the starting alcohol.

467. B Ethanol produces a greater drying effect which indicates that more must be evaporating. Evaporation is boiling.

468. B The number of moles indicates that there is only enough electrophile (1-pentanol) to react with the stronger nucleophile. Bromide is a better nucleophile, so it will react to give the major product.

469. A The hydroxide group is a terrible leaving group, so it must be protonated to form water, a good leaving group, before it will leave.

470. A The mechanism is S_N2 because the alcohol is primary and the nucleophile is strong. Answer C and D are eliminated because no double bond is formed.

471. A Sulfur is similar to oxygen so the thiol is similar to methanol.

472. C Potassium permaganate ($KMnO_4$) is an oxidant. Answer C is the only product in a higher oxidation state (more oxygens) than the starting material.

473. C The only thing that changes between the two mechanisms is the nucleophile, eliminating answers A and B. Methoxide is a stronger nucleophile because it is electron rich.

474. C The information given indicates that the mechanism will be S_N1 which produces a racemic mixture of I and II.

475. B The information given indicates that the mechanism will be S_N2 which produces a product of inverted stereochemistry (answer II).

476. D The information indicates that the mechanism is S_N1, so a tertiary haloalkane will react faster because of the more stable cation. Answer D is the only tertiary haloalkane.

477. A The information indicates that the mechanism is S_N2, so a primary haloalkane will react faster because of less steric hindrance. Answer A is the only primary haloalkane.

478. A Answer A is NOT true because it is a S_N1 mechanism. The product will be a racemic mixture.

479. B Answer B is NOT true because it is a S_N2 mechanism, so an optically active product results.

480. D The reaction mixture becomes homogenous because the product is soluble in water. Alcohols have a higher solubility in water than haloalkanes.

481. B Hydrogenation is always a reduction.

482. C The hydroxyl group labeled 3 has the least steric hindrance so it would be most likely to undergo a S_N2 reaction.

483. C Potassium chromate (K_2CrO_4) is an oxidant that can convert secondary alcohols to ketones.

484. C The carbon labeled 4 is a carboxylic acid which is the highest oxidation state in sodium cholate.

485. B The hydroxy groups cannot be converted to aldehydes because all three are secondary alcohols which can only be converted to ketones. Only primary alcohols can be oxidized to aldehydes.

486. B Potassium chromate is a strong oxidant and will oxidize the primary alcohol and aldehyde to carboxylic acids and the secondary alcohol to a ketone.

487. A $LiAlH_4$ is a reducing agent so product B must be at a lower oxidation state than product A.

488. A The K_2CrO_7 gains electrons because it is an oxidizing agent (i.e., it is reduced). Remember, when a compound is oxidized, it loses electrons, and when it is reduced, it gains electrons.

489. C Product A has two aldehydes and a ketone. One of the aldehydes and the ketone are products of the ozonolysis of the double bond.

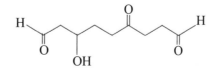

490. D Product A has a ketone, two aldehydes, and a hydroxyl group. The LiAlH$_4$ reduces them all to hydroxyl groups.
491. B The bromide ion attacks the electrophile in a substitution reaction.
492. D The OH must be protonated before it can leave as water.
493. B The product is shown below and the chiral center is circled.

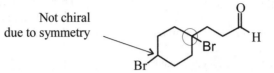

Not chiral
due to symmetry

494. C The dichromate will oxidize the aldehyde to a carboxylic acid and the hydroxyl group to a ketone.
495. A The addition begins with the protonation of the double bond (electrophilic addition) and then attack at the most substituted carbon (Markovnikov).
496. A Lithium aluminum hydride (LiAlH$_4$) is a reducing agent that will reduce the carboxylic acid and the ketone to alcohols.

497. C   An aldehyde has a hydrogen attached to a carbonyl carbon.

498. B   A ketal has two ether groups attached to a secondary carbon.

499. C   A hemiacetal has one ether group and one hydroxyl group attached to a primary carbon.

500. D   A hemiketal has one ether group and one hydroxyl group attached to a secondary carbon.

501. C   Answer C is not true because the carbonyl carbon is an electrophile (electron deficient), therefore it is subject to nucleophilic attack.

502. C   The carbonyl is planar so the carbon, oxygen and neighboring carbons are all in the same plane. The other answers include atoms that cannot be in the same plane due to the tetrahedral geometry of carbon atoms with four sigma bonds.

503. D   Pentanal is the most acidic because it is an aldehyde. The hydrogen attached to the α-carbon is more acidic than other hydrogens. Aldehydes are more acidic than ketones, but both are less acidic than alcohols.

504. B   The hydrogen attached to the α-carbon is the most acidic due to resonance stabilization of the anion formed by the carbonyl bond.

505. C   A beta hydrogen is converted to the carbon that is two carbons away from the carbonyl carbon.

506. D   A gamma hydrogen is converted to the carbon that is three carbons away from the carbonyl carbon.

507. C   The carbon labeled 3 has the most resonance stabilization of the conjugate base anion. Increased stability of a conjugate base leads to increased acidity. See the resonance structures below.

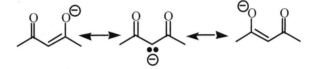

508. B   The carbonyl carbon is electrophilic because of its partial positive charge which results from a resonance structure of the double bond to the oxygen.

509. C   The carbon labeled 3 has a lone pair of electrons and a negative charge. It is very electron rich leading to increased nucleophilicity.

510. D   Answer D is the only structure in which that only electrons have been moved. Answers B and C shit hydrogens and answer A adds a double bond.

511. C   If a compound is more acidic, it must be due to stabilization of the conjugate base anion. The double bond in the ketone allows resonance stabilization of the alpha carbon. See the resonance structures in answer 507.

512. A   The net dipole has a partial positive charge on the carbon and negative charge on the oxygen due to resonance shown below.

513. C   Tautomers are formed when a hydrogen is moved within a structure. The shifted hydrogen is circled below.

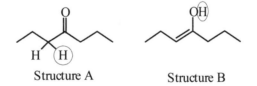

Structure A   Structure B

514. B   The tautomer with the ketone functionality is called the keto form.

515. A   The tautomer with the double bond (ene) and hydroxyl (ol) is called the enol form.

516. A   Alcohols have the highest boiling points due to hydrogen bonding. Although ketones and aldehydes can hydrogen bond with water, they do not hydrogen bond with themselves.

517. A   Since alcohols have the greatest ability to hydrogen bond, their water solubility is the highest (like dissolves like).

518. D   Ketones follow the same trend as alkanes. As chain length gets longer (i.e., molecular weight increases), the boiling point increases.

519. A   Shorter chain-length (nonpolar portion) leads to increased water solubility.

520. C   The solution would be a 5% solution. Since 2-pentanone would be soluble based on the information in the question, 2-butanone will be completely soluble. This is true because it has a shorter chain-length.

521. C   The lower the molecular weight of a ketone, the more soluble in water. Ketones are able to hydrogen bond with water, but not with themselves.

522. A   Because methanal will have a lower boiling point then ethanal, it is in the gas phase at room temperature.

523. D Redox reactions occur in pairs, so the reduction of the silver must be coupled to an oxidation. Based on the reactivity of aldehydes, they are able to oxidize to carboxylic acids.

524. A Aldehydes react with the Tollen's reagent and pentanal is the only aldehyde.

525. B The butanal is oxidized to a carboxylic acid and the silver ion is reduced to metallic silver.

526. B Methanol, a nucleophile, will attack the carbonyl of butanal to form a hemiacetal. A second methanol molecule will do another nucleophilic attack on the hemiacetal to form the acetal.

527. B The oxidation of a primary alcohol to aldehyde requires a mild oxidant. If K_2CrO_4 is used, it will oxidize the alcohol to a carboxylic acid.

528. A The addition of OH and H to a double bond is called hydration (the addition of water).

529. C The product is formed by moving a single hydrogen.

530. B The intermediate has an alkene (en) and a hydroxyl group (ol).

531. A The hydration of the alkyne will always place the hydroxyl group on the most substituted carbon (Markovnikov), so an aldehyde cannot be formed.

532. B The hydration would produce the Markovnikov alcohol (most substituted) and there would be no double bond leading to tautomerization.

533. C The transformation consists of one pi bond exchanged for two sigma bonds. This is an addition, eliminating answers B and D. The first step is the nucelophilic attack of the water on the aldehyde.

534. C Hydroxide is a much stronger nucleophile than water.

535. A The carbonyl carbon has a deficiency of electrons (a partial positive charge) so it is the electrophile.

536. A The acid protonates the oxygen of the carbonyl which draws more electron density away from the carbonyl carbon. This carbon becomes more electrophilic as it becomes more electron deficient.

537. D In an acidic solution, hydroxide ion is not present. Therefore, water functions as the nucleophile and attacks the activated electrophile.

538. A Answer C and D are untrue, leaving answer A and B. Based on the question, the 2-butanone is less reactive which means it is less electrophilic (electron deficient). To account for the difference in reactivity, alkyl groups must be electron donating making 2-butanone's carbonyl carbon less electrophilic.

539. B Two molecules of methanol would add to the formaldehyde forming an acetal.

540. C Sodium chromate (Na_2CrO_4) oxidizes a secondary alcohol to a ketone.

541. A The reaction is the reverse of oxidation (reduction) so the reducing agent $NaBH_4$ will achieve the conversion.

542. B The nucleophile must contain a CN group eliminating answers C and D. HCN is too weak of a nucleophile for a reaction to take place. It would have to be CN⁻ that is the actual nucleophile.

543. A The Grignard reagent must be nucleophilic because it attacks the electrophilic carbon of a carbonyl, eliminating answers C and D. The Grignard reagent is a very strong nucleophile.

544. C Diethyl ether is the only solvent that a strong nucleophile does not attack. Water and ethanol both have acidic hydrogens that are removed by the nucleophile (Lewis base) and dichloromethane would be subject to an S_N2 displacement.

545. B The product of the reaction is shown below. The chiral center is circled; however, the product would not be optically active. There would be a racemic mixture since the carbonyl is planar and can be attacked from both sides.

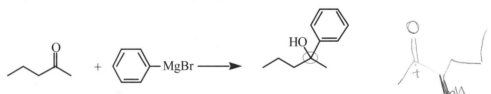

546. C The reaction converts one pi bond to two sigma bonds, so it is an addition. The reaction mechanism begins with an attack by the Grignard reagent (a nucleophile).

547. C Answer C is the only structure that contains a carbon-magnesium-halogen bond.

548. C In an aldol addition one of the carbonyls is converted to an enolate ion and the alpha carbon of the enolate ion attacks the other carbonyl. The attacked carbonyl becomes an alcohol and the enolate carbonyl remains a carbonyl.

549. D The enolate ion is the nucleophile because it is the electron rich substance.

550. C Because the reaction is base catalyzed, the first step cannot be protonation, eliminating answers A and B. The deprotonation forms the enolate which is the nucleophile.

551. A In an acid catalyzed reaction, the acid will protonate the carbonyl oxygen making the carbonyl carbon more electron deficient thus activating the electrophile. In a base catalyzed reaction, the base deprotonates the alpha carbon activating the enolate nucleophile.

552. B Dehydration (removal of an OH and H) of the alcohol component of an aldol product gives an α-β unsaturated carbonyl.

553. D Ethanal can react with itself or with propanal and act as the nucleophile. Alternatively, ethanal can react with itself or propanal while propanal acts as the nucleophile. Each of the these produces a different structural isomer.

554. A Answer A is the only structure with one carbonyl and one hydroxyl group so the answer is easy to identify. However, the mechanism for the product is shown below.

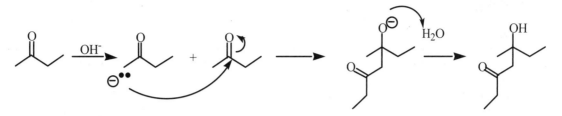

555. A The aldehyde in answer A is the only one with no alpha hydrogens so that it cannot form an enolate ion and it must be the electrophile.

556. D In the first step of the reaction, the KH forms the enolate ion, a strong nucleophile, so that an S_N2 reaction occurs with CH_3I in the second step. <u>CH_3I is not a nucleophile</u> so it will not attack the carbonyl.

557. B As described in answer 556, the reaction is a nucleophilic substitution.

558. A The methyl iodide is attacked by the nucleophile (the enolate ion), therefore it is the electrophile.

559. C Both the carbonyl carbon (1) and the alkene carbon (3) are electrophilic. As a result, 1,2 and 1,4 additions occur.

560. B The product is the result of the <u>nucleophile attacking</u> the alkene and forming an enol that tautomerizes back to a ketone.

561. D The arrow in the diagram points to the bond that is broken. The bond can be formed from the products if the alpha hydrogen on the product on the left is removed to form an enolate and it attacks the carbonyl on the product on the right.

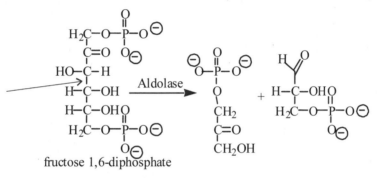

fructose 1,6-diphosphate

562. B The product results from the dehydration of the aldol product. The alkene is between the alpha and beta carbons.

563. A The reaction is an aldol condensation because it combines two ketones to form an alcohol and a ketone. It is base catalyzed because hydroxide, a base, is added to the reaction mixture.

564. C The formation of the product results from the removal of OH and H from the intermediate.

565. A The mechanism for the reaction is shown on the next page. The enolate ion is formed by removing the hydrogen from carbon 1.

566. C The mechanism for the reaction is shown on the next page. The carbon that is attacked by the nucleophile is the electrophile. The carbon attacked is carbon 3.

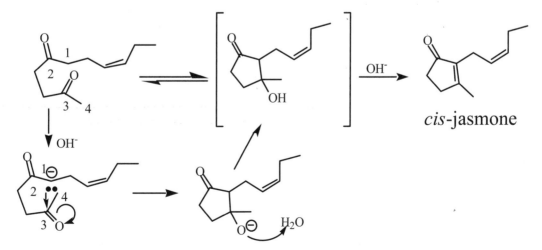

cis-jasmone

567. A   The leaving group would be CH_3^- before the substitutions of the I atoms for hydrogens. After the substitutions the leaving group is CI_3^-. CI_3^- is a better leaving group because the I atoms are electron withdrawing and stabilize the negative charge.

568. C   The reaction requires a basic solution so the reaction needs to be aqueous. Dichloromethane and acetone would both react with iodine.

569. A   Only structure I is a methyl ketone and the information given about the test states that it only works for methyl ketones.

570. A   The reaction is an aldol condesation followed by a dehydration. It combines an aldehyde and a ketone to give an alpha-beta unsaturated ketone. It is base catalyzed because OH^- is present in the product.

571. A   Acetone is the only carbonyl with an alpha hydrogen, so it forms the enolate ion and acts as the nucleophile.

572. B   Benzaldehyde is the electrophile because it has no alpha hydrogen to form the enolate ion. Instead, it is attacked by the enolate ion of acetone.

573. C   There is only one product because only one enolate ion forms. The symmetry of acetone (II) means that the removal of either alpha hydrogen produces a single enolate ion. The fact that the benzaldehyde has no alpha hydrogen (I) means that it cannot form an enolate ion.

574. A   (S)-3-methyl-2-heptanone would lose its chirality because the chiral carbon is the alpha carbon. It is the configuration of the alpha carbon that is affected by tautomerization.

575. C   Product A is a ketal because it has two ethers attached to a secondary carbon.

576. A   The diol is a nucleophile that attacks the carbonyl carbon twice.

577. C   The diol oxygens are both substituted for the carbonyl oxygen.

578. C   The $Na_2Cr_2O_7$ oxidizes the alcohol in product C to form a new ketone in addition to the one already present in product C.

579. A   2,2-dimethylbutanal cannot form an enolate ion beause it does not have any alpha hydrogens. Only product D can form an enolate. The alpha hydrogens between the two ketones are significantly more acidic than the alpha hydrogens on the terminal carbons, so only a single enolate ion will form.

580. C   The treatment of product D with $LiAlH_4$ will produce a diol that can form a ketal when reacted with a ketone.

581. B   The first step of the reaction protects the ketone from nucleophilic attack by decreasing the electrophilicity of the carbonyl carbon.

582. C   The product formed is shown below.

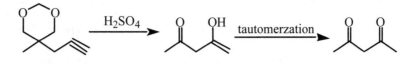

583. A   Product A is shown below containing zero carbonyls. The nucleophile attacks the carbonyl and then the oxygen is protonated.

584. B   Product C is shown below. It contains one carbonyl.

585. C   The new product is shown below. It contains two carbonyls.

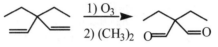

Product B

586. B   The oxygen comes from the ozonolysis of the alkene. The oxygen comes from O_3. See answer to question 584.
587. B   Product D is shown below. It is an α,β-unsaturated carbonyl.

588. B   The transformation of product C to product D is a dehydration of an alcohol to form an alkene.
589. C   Product D is an aldehyde which will be oxidized by $Na_2Cr_2O_7$ to produce a carboxylic acid.
590. C   Hydrogenation will remove the double bond of product D.
591. A   The nucleophile (CN^-) attacks the electrophilic carbon which is beta to the carbonyl. The hydrogen adds to the carbon at the alpha position. Since, the two additions are on adjacent carbons, they constitute a 1,2 addition.
592. D   Formaldehyde does not have an alpha hydrogen so no enolate ion can form.
593. B   The structure in answer B requires the movement of hydrogens.
594. A   The diene attacks the dienophile.
595. D   The dienophile is an α–β unsaturated carbonyl. Recall that nucleophiles attack α,β unsaturated carbonyls at the β-carbon, carbon 4 in this case.
596. B   The two sites of addition are separated by two carbon atoms.
597. A   The structure in answer A has no alpha hydrogens to form an α,β-unsaturated carbonyl through dehydration.
598. D   Carboxylic acids follow the same trends as alkanes. Increasing the length of the carbon chain increases the boiling point.
599. B   Carboxylic acids form stronger hydrogen bonds than alcohols, so the carboxylic acid has the highest boiling point.
600. B   Chain length has nothing to do with hydrogen bonding, so answers A and C are eliminated. The larger the molecular weight, the stronger the London dispersion forces.
601. B   From the chart, we see that the unknown acid has to have a chain length shorter than heptanoic acid and longer than butanoic acid. Answer B (hexanoic) is the only carboxylic acid that fits this requirement.
602. A   Carboxylic acids form the strongest hydrogen bonds, so they have the greatest ability to bond with water molecules. This gives them the highest water solubility.
603. C   The proponaic acid is more soluble than the octanoic indicating increased solubility, eliminating answers B and D. The shorter the alkyl chain, the higher the solubility of the acid in water.
604. A   Carboxylic acids get their names from the fact that they have highly acidic protons.
605. D   Nitro groups are electron withdrawing which makes the proton more acidic by stabilizing the conjugate base anion.
606. C   Acid chlorides have more acidic hydrogens than ketones, aldehydes, or esters.
607. D   Nitro groups are electron withdrawing, eliminating answer A. Nitrobenzoic acid has the lower pK_a indicating that it is more acidic, so it must stabilize the negative charge of the conjugate base.
608. B   The difference between the two structures is the carbonyl group in the acid which resonance-stabilizes the negative charge of the conjugate base.
609. D   The structure in answer D has the lowest pK_a therefore it has the most acidic proton.

610. C Since the electron withdrawing group is closer to the acidic hydrogen, it has a greater stabilizing effect.

611. C According to the table, an addition of a Cl atom lowers the pK_a of an acid. Thus, the pK_a of 2,2-dichloroethanoic acid will be between one Cl atom and three Cl atoms. Answer C is the only value that fits this criteriaon.

612. D Both carbon oxygen bonds are involved in the resonance of the negative charge. The bond length is between the single and double bond.

613. A An acid chloride is more reactive than an ester so the alcohol will combine with the acid chloride to form an ester by displacing the chloride atom.

614. C Similar to the alcohol formation of an ester, an amine combined with a carboxylic acid forms an amide.

615. A Alcohols are oxidized to carboxylic acids.

616. C Thionyl chloride ($SOCl_2$) is the reagent that converts butanoic acid to butyl chloride because it provides a nucleophilic chloride atom.

617. A An acid chloride plus a carboxylate ion forms an anhydride. The carboxylic acid, ester and amide are more stable than the anhydride, so they will not react.

618. C Ethanol will react to displace the OH and form the ethyl ester. An alcohol and an acid combine to form an ester.

619. B The oxygen in ethanol is nucleophilic and attacks the carbonyl carbon. The oxygen in the hydroxide group of the acid leaves as water.

620. C Beta keto carboxylic acids can decarboxylate releasing CO_2.

621. C An anhydride has an oxygen atom between two carbonyl carbons.

622. B The anhydride is formed because it is more reactive than the acid, which enhances the reaction with hydrazine.

623. C Hydrazine is a nucleophile and sigma bonds are substituted for sigma bonds.

624. B Fewer bonds to oxygen are present in the product than in the reactant.

625. D All three will increase the formation of an ester. Increasing the concentration of alcohol increases the concentration of a reactant. Anhydrides are more reactive than carboxylic acids leading to more product. Acid catalyzes the reaction.

626. D The same product is formed in each reaction so answers A and B are eliminated. Acid chlorides are more reactive the carboxylic acids so they do not require a catalyst.

627. C The acid protonates the OH of the carboxylic acid converting it to water, a good leaving group. It also activates the electrophile by protonating the carbonyl oxygen, which makes the carbonyl carbon more electron deficient.

628. C The Grignard reagent is a strong nucleophile that replace one sigma bond with another sigma bond; therefore, the reaction is a nucleophilic substitution.

629. A The conjugation is not stabilizing any charges so it has nothing to do with the relative reactivities, eliminating answers C and D. Benzophenone must be more reactive causing the Grignard reagent to preferentially react with the benzophenone over the methyl benzoate.

630. A The tertiary alcohol cannot be oxidized to any carbonyl, so all that can be done is removal of the proton by a base.

631. D Both benzophenone and methyl benzoate do not have acidic protons, eliminating answers B and C. Triphenylcarbinol is more acidic than ethanol because it has extensive resonance stabilization of the conjugate base's negative charge.

632. A The acetate ion is electron rich so it is a nucleophile.

633. A The ester will be optically active because the reaction will be an S_N2 mechanism with inversion of stereochemistry.

634. B The acetate ion can also be a base and remove a hydrogen atom to start a dehydrohalogenation.

635. D Because the reaction is a S_N2 reaction, the rate determining step depends on the concentration of both the electrophile and the nucleophile (I and II). Raising the temperature almost always speeds up a reaction (III).

636. B The acid chloride will be more reactive than the carboxylic acid, eliminating answers C and D. The amide has two ethyl groups attached, so diethyl amine is required.

637. B Beta keto esters are subject to decarboxylation reactions, and answer B is the only beta keto ester.

638. A The first step of the reaction scheme is necessary to activate the electrophile by creating a more reactive species.

639. B The anilide has an amine group next to a carbonyl making it an amide.

640. C The SO_2Cl provides a strong nucleophilic chloride so that the chloride ion can substitute for the hydroxide group.

641. B The lone pair of electrons on the nitrogen atom is nucleophilic and attacks the electrophilic carbonyl carbon.

642. C The nucleophilic amine attacks the carbonyl carbon and replaces the chloride atom.

643. C The bromine atom is electron withdrawing making the carbonyl carbon more electrophilic.

644. B There is no amide product eliminating answer A. The triethylamine contains a proton in the product indicating that it has functioned as a base.

645. C Resonance stabilization of the conjugate base anion, increases the acidity of the phenol hydrogen (II). The chlorine atoms are electron withdrawing also stabilizing the negative charge (I).

646. B The oxygens next to the carbonyls are attached to alkyl groups.

647. A The hydrogen is removed from the phenol to make an electron rich anion that can attack the electrophilic carbonyl.

648. B The name indicates the plane of symmetry that is in the middle of the molecule.

649. B The intermediate is a peroxide with two oxygen atoms linked in a sigma bond. The presence of the extra oxygen indicates a higher oxidation state. Also hydrogen peroxide is an oxidizing agent.

650. B Decarboxylation indicates that the reactant loses CO_2, so answers A and C are eliminated. The product is a Z alkene so the phenyl groups need to be on the same side, answer B.

651. B A ketone is formed in the place of a carboxylic acid. The ketone is at a lower oxidation state.

652. C The name indicates that the benzoyl group is a 2 substituent on benzoic acid. The group is circled below.

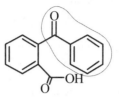

653. B The carbonyl is the electrophile in the reaction.

654. C The acid provides a proton to turn the hydroxide into a good leaving group, then the proton is regenerated in the last step of the reaction.

655. C Based on the generic reaction, the OR' attaches to the carbonyl which would give the product in answer C.

656. A The reaction is driven toward products by an excess of the nucleophile, (i.e., methanol). The addition of water just adds another nucleophile which can compete with methanol.

657. D Hydroxide is not a good leaving group, so it must be protonated to form water before it can leave.

658. B An alkoxide is not a good leaving group, so it must be protonated to form the original alcohol before it can leave.

659. C The intermediate is the enolate ion of the original ketone which results from tautomerization and then deprotonation.

660. B The formation of two caboxylic acid groups from a ketone is the result of oxidation. Carboxylic acids are at a higher oxidation state than ketones.

661. D It has two carboxylic acid groups.

662. A A diol will result from the reduction of the carboyxlic acid groups.

663. B Methanol will do a nucleophilic substitution to form an ester. Alcohols and carboxylic acids combine to form esters.

664. D Without the base, no tautomerization will occur. A ketone cannot be oxidized further.

665. B Nylon 6.6 has nitrogen atoms next to carbonyls, indicating an amide. The brackets with the subscript n denote a repeating group.

666. A The lone pair of electrons on the nitrogen of the amine attacks the electrohpilic carbonyl carbon of the acid.

667. B Amides are more stable (less reactive) than esters. Dacron is a polyester.

668. D The pyridine is protanted in the product, so it has reacted as a base.

669. B Carboxylic acids have protons that are more acidic than alcohols, so the base would deprotonate the acid instead of activating the nucleophile.

670. A The carbonyl is more reactive in 3,5-dinitrobenzoyl chloride because it is an acid chloride (which is more reactive than an ester). Additionally, and it has electron withdrawing groups on the ring making the carbon even more electrophilic.

671. D The nitro groups are electron withdrawing, so they activate the electrophilic carbonyl carbon by withdrawing electron density.

672. C The substitution takes place at the carbonyl not the chiral carbon, so the chirality is not affected.

673. A Phenolphthalein has an oxygen attached to an alkyl group and a carbonyl at low pH (i.e. acidic conditions before NaOH, a base, is added).

674. A The hydrolysis of the ester forms an alcohol and a carboxylic acid that is deprotonated by the base.

675. B The molecule loses water to form an alkene.

676. C Phenolphthalein is a carboxylic acid that is deprotonated.

677. B The conversion of an ester to a carboxylic acid results from hydrolysis.

678. A Maleic acid and fumaric acid are geometric isomers of each other, therefore, one is formed from the other by isomerization.

679. C The water molecule is the nucleophile. The lone pair of electrons on the oxygen attacks the electrophilic carbonyl carbon.

680. A They have the same connectivity and different spatial orientations, but are not mirror images. Geometric isomers are diastereomers.

681. B Fumaric acid is more stable because the *trans* isomer has less steric hindrance.

682. D The $LiAlH_4$ would form a diol by reducing the carboxylic acid all the way to the alcohol.

683. D Two moles of the product shown below are produced by the ozonolysis of maleic acid.

684. A An anhydride is more reactive than a carboxylic acid.

685. B As the CO_2 bubbles out of solution, there is no reactant remaining to do the reverse reaction.

686. D Carbonyls on the beta carbon of a carboxylic acid catalyze the decarboxylation reaction.

687. A The acid protonates a carbonyl oxygen of the propionic anhydride. Electrophiles are activated by protonation and the anhydride is the electrophile, therefore, propionic anhydride must be protonated.

688. B Propionyl chloride is an acid chloride, which is the most reactive carboxylic acid derivative.

689. B According to the question the purpose of the base is to remove the excess anhydride, so it must hydrolyze the anhydride. The isoamyl alcohol will have been all used up in the reaction eliminating answer C.

690. A Esters can be hydrolyzed by a strong base and the product would be lost.

691. B Piperazine has nitrogen atoms with two alkyl groups attached, therefore, it is a secondary amine.

692. A Amphetamine contains a nitrogen atom with only one alkyl group attached, therefore, it is a primary amine.

693. D The ester and the quaternary amine (four alkyl groups attached, always positively charged) are circled below.

694. C The N,N in the name indicates two substituents on the nitrogen atom in addition to the regular chaingiving a total of three alkyl groups attached to the nitrogen.

695. D Answer D is the only amine with 2 methyl groups and on ethyl group attached to it.

696. C The name indicates that there are four carbons and the amines are on the first and fourth carbon. The structure in answer C is the only one that meets this criteria.

697. D The presence of the lone pair of electrons causes constant interchange of the stereochemistry resulting in an amine that is not optically active.

698. D The addition of the fourth substituent removes the lone pair of electrons and stops inversion.

699. B The amine on the right has a chiral center at one of the carbons attaching to the nitrogen.

700. A The dipole is in the direction of the lone pair of electrons.

701. C The triethylamine does not hydrogen bond because it does not have a N – H bond, while the other two amines (II and III) do contain such a bond.

702. D Although triethylamine cannot hydrogen bond with itself, it can hydrogen bond with water because of the presence of highly electronegative nitrogen.

703. A Ethylamine has two N-H bonds that can hydrogen bond with water, increasing its solubility. The other choices are tertiary amines which can not hydrogen bond.

704. A Cyclohexylamine is the only substance listed that can form hydrogen bonds on its own. It is the most polar, so it is the most soluble in water.

705. D Both of the amines have approximately the same molecular weight, eliminating answer A and B. The primary amine hydrogen bonds, and therefore has a much higher boiling point.

706. C Both the amine and alcohol have approximately the same molecular weight, eliminating answers A and B. The higher boiling point indicates that ethanol must have stronger hydrogen bonding. Oxygen is more electronegative making the hydrogen bonds in ethanol stronger.

707. D Triethyl amine has no hydrogen bonding.

708. A Amines are basic so they raise the pH of aqueous solutions.

709. B A Lewis base donates electrons.

710. A The lower pK_b value indicates that p-toluidine is more basic, eliminating answers B and D. The alkyl group is electron donating which makes the amine more electron rich and therefore more basic.

711. C In the name, N-methyl indicates that the methyl group is on the nitrogen atom. Answer C is the only choice with the methyl group on the nitrogen.

712. C The addition of electron donating alkyl groups to the nitrogen makes the amine more basic because it is richer in electron density.

713. C The trend in the table indicates that the pK_b of the bromoaniline must be in between chloro and iodo, so the value must be between 10.00 and 10.22. Answer C is the only value that meets this criteria.

714. A Methoxy is a stronger electron donating group than an alkyl group; therefore, the pK_b must be lower than toluidine and answer A is the only value lower than 8.92.

715. B The pK$_b$ for N-methylaniline must be between the value of aniline (9.40) and N,N-dimethylaniline (8.94). Answer B is the only value that meets this criterion.

716. D The reaction favors the reactants indicating that the reactants are more stable. The aromaticity of the reactant stabilizes it. The lone pair of electrons on the nitrogen atom plus the 4 electrons from the pi bonds meet the (2n + 2) rule for aromaticity.

717. C The pyrrole accepts a proton and donates electrons, therefore it is both a Bronsted-Lowry base and a Lewis base.

718. A The substitution of the ring indicates the ortho-para directing.

719. A NaHCO$_3$ is a base, so it cannot do the reactions stated in answer B, C, and D.

720. A The amine group activates the ring for substitution at the two ortho sites and the one para site.

721. C The amine will continue to perform a nucleophilic attack on methyl iodide until a quaternary amine is formed and there is no lone pair of electrons present.

722. B The diethylamine already has two alkyl groups, so two more alkyl groups add to form the quaternary amine.

723. A By adding the methyl iodide slowly to the large excess of ethyl amine, the methyl iodide is kept at a low concentration preventing substituted amines from reacting with it.

724. B The ethylamine is electron rich and attacks the electrophilic methyl iodide.

725. D The excess methyl iodide ensures that the reaction continues until no lone pair of electrons is present on the nitrogen of the amine. Only a quaternary amine has no lone pair of electrons.

726. A Primary amines react with ketones to form an imine.

727. B Secondary amines react with ketones to form enamines.

728. B Enamines have a double bond attached to the same carbon as the nitrogen.

729. A A primary amine will form an imine when reacted with the ketone.

730. B The mechanism is shown below.

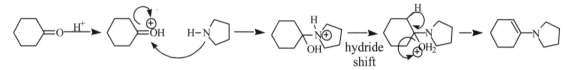

731. B See the mechanism in the answer to question 730. The electrons on the nitrogen atom attack the electrophilic carbonyl carbon.

732. B Sulfur is similar to carbon so the sulfonyl chloride is similar to an acid chloride.

733. A The hydrogen on the nitrogen is acidic. In the basic reaction solution, the product is deprotonated to form a salt which is soluble in water.

734. A The product is analogous to an amide with a nitrogen attached to a sulfur that is double bonded to oxygen.

735. D Tertiary amines will form a quaternary amine that does not have a hydrogen which can be removed in order to neutralize the positive charge.

736. C In dimethyl amine, both R$_1$ and R$_2$ are methyl groups.

737. A The electrophilic sulfur atom is attacked by the nucleophilic amine.

738. C The structure of benzoyl chloride is shown below. The nucleophilic nitrogen atom will attack the electrophilic carbonyl carbon and will substitute for the chloride ion.

benzoyl chloride

739. A The benzesulfonamide has two oxygens that can resonance stabilize the conjugate base's negative charge. The benzamide has only one oxygen to resonance stabilize the anion.

740. C An amide has a nitrogen atom attached to a carbonyl carbon.

741. B The carbonyl group is electron withdrawing making the amide less basic than the amine.

742. D The amide is not reactive with nitric acid, but the amine would react with the acid.

743. B The substitution is para, so the amide must be ortho, para directing. The best explanation for the lack of ortho substitution is steric hindrance.

744. C The ring would be deactivated because the group would become electron withdrawing. All electron withdrawing groups are meta-directing except for halogens, which are ortho-para directing.

745. C An addition of an alkyl group to an amine is called an alkylation.

746. D The nitrogen atom has four substituents and it is positively charged.

747. A The nitro groups are electron withdrawing which stabilizes the negative charge making the phenol more acidic.

748. C   Only tertiary amines form amine picrates, and answer C is the only tertiary amine.

749. B   The nucleophilic amine displaces the iodide atom to form the new product.

750. A   The electrostatic forces between the salt increases the intermolecular forces raising the boiling point and making the salt less volatile. Consequently, there is less odor. The other answer choices can be ruled out because there is a negligible difference in molecular weight (answer D) and ephedrine is a secondary amine (answer C).

751. D   There are two alkyl groups attached to the nitrogen.

752. A   A nitrogen with a double bond indicates an imine.

753. B   An amine has fewer oxygens so it has a lower oxidation state than an oxime. $LiAlH_4$ is a reducing agent.

754. A   Benzylamine is a primary amine and answer A is the only imine which is the product when a primary amine is reacted with a ketone.

755. A   The reaction scheme shows that NH_2OH produces primary amines when reacted with ketones.

756. B   The reaction scheme shows that primary amines produce secondary amines when reacted with ketones.

757. D   The product is formed by moving a hydrogen so it is a tautomerization.

758. B   A double bond is attached to the same carbon as nitrogen producing and enamine.

759. B   A double bond is attached to the same carbon as nitrogen producing and enamine.

760. A   A new sigma bond is formed replacing a hydrogen with a benzyl group; therefore, it is a substitution.

761. A   The nitrogen atom has three alkyl groups attached to the nitrogen, indicating a tertiary amine.

762. A   Answer A is the only answer that makes sense. Answer D has nothing to do with the question. Answer C is the opposite of the truth. If a more stable product were formed it would be less reactive, which would defeat the purpose of the reaction. There is no stereoselectivity in the reaction.

763. A   An amide is at a higher oxidation state than an amine.

764. B   The cyano group has a carbon atom triple bonded to a nitrogen atom. The product has a single bond between carbon and nitrogen. The reagent must provide a source of hydrogen atoms to reduce the triple bond to a single bond. Answer B is the only reducing agent.

765. A   An oxime is at a higher oxidation state than an amine, so an oxidizing agent must be used. Answer A is the only oxidizing agent.

766. B   An amine is at a lower oxidation state than a nitro group, so a reducing agent must be used. Answer B is the only reducing agent.

767. C Fatty acids are alkane chains with a carboxylic acid attached to one end.
768. B Amino acids contain both an amine group and a carboxylic acid.
769. C Section 1 is the long alkane chain so it is nonpolar and hydrophobic, while section 2 is a carboxylic acid so it is polar and hydrophilic.
770. A The side chain is the group attached to the alpha carbon of the amino acid (circled below).

771. C The amino acid in the question is rotated from its regular perspective. The amine group facing down is actually part of the backbone. Below is the same amino acid in the more familiar configuration.

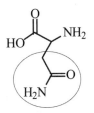

772. D The amino acid is considered nonpolar because its side chain is an alkane which is nonpolar.
773. C The amino acid is considered polar because its side chain is a short alcohol. The polar hydroxide group makes the amino acid polar. When grouping amino acids into acid-base and polar-nonpolar categories, it is the side chain that is the determining factor.
774. B The amino acid is considered polar because its side chain contains an amine. The amine group also makes the amino acid basic because it functions as a base.
775. A The side chain of the amino acid contains a carboxylic acid making the amino acid both polar and acidic.
776. C Amino acids are the building blocks of proteins.
777. D Essential means that the amino acid cannot be synthesized in the body. These amino acids are *essential* in the human diet.
778. A An increase in the length of the fatty acid's carbon chain leads to a decrease in the polarity and thus a decrease in water solubility.
779. D Two amino acids linked by a amide linkage (peptide bond) are called dipeptides.
780. C Two fatty acids linked by an ester bond is called a diglyceride.
781. D The structure has three ester functional groups. A triglycerol was used to make the triester (triglyceride).
782. A The reaction adds an OH and H across a cleaved bond making it a hydrolysis.
783. B Lipases are used to cleave the ester bonds of triglycerides in the human body.
784. D Sodium hydroxide will achieve ester hydrolysis.
785. D The hydrolysis of triglycerides in fat with NaOH is called saponification.
786. B Stearic acid is not amphoteric (able to act as an acid or a base). It cannot function as a base.
787. A Adipose cells are where triglycerides are stored.
788. D Increased levels of epinephrine start the lipolysis process.

789. C   The structure is a dipeptide. Each amino acid is circled below.

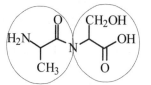

790. C   The amide bond has partial double bond character preventing rotation around the amide linkage (III is eliminated) and the hydrogen bonding between side chains also helps to determine secondary structure.

791. B   The structure of valine is not needed to answer this question. When the pH is equal to the pI, the overall amino acid is neutral, but it exists as a zwitterion (a dipolar ion).

792. D   When the pH of the solution is greater than the pI, the amino acid has a negative charge and the protonated amine is deprotonated.

793. A   When the pH of the solution is less than the pI, the amino acid has a positive charge and both ends of the molecule are protonated.

794. C   Both fatty acids have the same chain length so the molecular weights are very close, eliminating answers A and B. The *cis* double bonds make it difficult for the chains to stack and therefore the intermolecular forces are weakened, lowering the melting point.

795. A   Oleic acid is an unsaturated fat because it contains a double bond.

796. C   Lauric acid has the shortest chain (hydrophobic region) so it will be the most soluble in water.

797. A   Oleic acid is the only fatty acid in the list with a melting point lower than room temperature, therefore, it is the only oil.

798. C   Linoleic acid has two cis double bonds which will put the melting point in the middle of oleic acid (4°C) and linolenic acid (-11°C). Answer C is the only value that falls in between these two values.

799. A   The longer the chain, the higher the melting point. This trend is supported by the table. Arachidic acid has 2 more carbons than stearic acid which has a melting point of 70°C. Answer A is the only answer that is greater than 70.

800. A   Both fatty acids have the same number of carboxylic acid groups, so they do the same amount of hydrogen bonding, eliminating answers C and D. Palmitic acid has more surface area because its carbon chain is longer, therefore its London dispersion forces are stronger.

801. A   The table indicates that the presence of double bonds lowers the melting point creating oils. Hydrogenation removes double bonds raising the melting point and creating solid vegetable oil to replace the solid lard.

802. D   Stearic acid contains the longest chain (the most carbons), therefore the β-oxidation cycle will run the most times for this fatty acid producing the most energy.

803. C   Lauric acid contains the shortest chain (the least carbons), therefore the β-oxidation cycle will run the least times for this fatty acid producing the least energy.

804. C   Eleostearic acid and linolenic acid have the same number of degrees of unsaturation eliminating answers A and B. The melting point is higher because two of the bonds in eleostearic acid are trans helping packing rather than hindering it.

805. D   Eleostearic and oleic acid have the same number of carbons leading to similar molecular weights, eliminating answers A and B. Both fatty acids have one *cis* bond, so they have the same number of "kinks", eliminating answer C. Packing is aided by the rigid *trans* structure raising the melting point.

806. D   Stearic acid has the longest chain therefore the micelles formed would have the greatest diameter.

807. A   Oleic acid is the only fatty acid of the choices that is an oil at room temperature.

808. B   One equivalent of NaOH would be required to remove the proton on the carboxylic acid forming the neutral zwitterion.

809. A   The fact that the amine group is protonated preferentially over the carboxylate ion indicates that the amine group is a stronger base.

810. A   Arginine is a basic amino acid, therefore, the pI value is greater than 7.

811. C   Aspartic acid is an acidic amino acid, therefore, the pI value is less than 7.

812. A   The three-fold channel is lined with polar amino acids, therefore, the channel will be hydrophilic. However, the four-fold channel is lined with nonpolar amino acids, therefore these channels are hydrophobic.

813. A   Fe is charged therefore the channel it uses must be polar in order for the ion to be soluble inside it.

814. A   The amino acid in answer A is the only alpha amino acid shown. All amino acids in the body are alpha amino acids.

815. C   The amino acid in answer C is a beta amino acid. Only alpha amino acids are used to synthesize proteins in the human body.

816. C   The nonpolar chains of the fatty acids are soluble in the grease and the polar heads of the fatty acid are on the surface of the grease to make it soluble in water.

817. C   The phosphoglyceride has two anions that must be neutralized by HCl.

818. B   The head has the negative charge making it polar, while the tail long long alkyl chains which are nonpolar.

819. B The heads must face the surface creating the hydrophilic surface. The tails face each other creating the hydrophobic interior.

820. D Histidine contains amine groups on the side chain making it basic and polar.

821. A The nitrogen in histidine has a lone pair of electrons that form the bond with iron. The amine is the Lewis base and the Fe is a Lewis acid.

822. D Since histidine is a basic amino acid, the pI must be greater than 7. Answer D is the only answer greater than 7.

823. B The pI is the pH value at the first equivalence point. See the graph below.

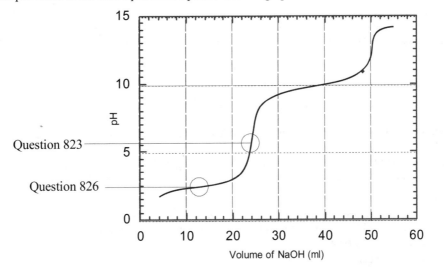

824. B There are two bumps in the titration curve indicating a diprotic acid.

825. B The zwitterion is the neutral double charge molecule that represents 100% of the species at the pI. See the answer to question 823.

826. A The molecule is 50% positive and 50% neutral at 1/2 way to the first equivalence point. See the titration in the answer to question 823.

827. D At a pH of 12.5 all acids on glycine are deprotonated giving an overall negative charge on the molecule.

828. B At a pH of 6 the molecule is in the zwitterion form.

829. A At a pH of 1.5 the overall molecule has a positive charge.

830. C The carboxylic acid proton is more acidic than the amino group protons. When the carboxylic acid proton is removed, the zwitter ion is formed.

831. B The priorities are arranged in a clockwise orientation, however the hydrogen is coming out of the page.

832. A D-serine has the opposite absolute configuration as L-serine.

833. A The structure in answer A is the only one that is a L amino acid. The body only contains L amino acids.

834. B The connection of amino acids forms a polypeptide.

835. B Proline is a secondary amine because it has the two alkyl groups connected to the nitrogen.

836. B In hydrolysis, water is added as a reactant break apart another molecule.

837. C Leucine is a nonpolar amino acid.

838. B The 20% HCl lowers the pH and protonates the amino acids.

839. A The amide bonds are converted to amines and carboxylic acid.

840. B Carboxylic acids are more reactive than amides.

841. C Aspartic acid is an acidic amino acid, therefore it has a pI less than 6 and will be negatively charged at pH 6, so it will migrate to the positive charge (right).

842. D Lysine is a basic amino acid, therefore its pI is greater than 6 and it will have a positive charge at pH 6 and migrate towards the negative charge (left).

843. C Apartic acid is an acidic amino acid; lysine is a basic and valine is neutral. Therefore they are easiest to separate because each will have a different charge at pH 6.

844. B If the pH is raised to 9.7, the lysine will be neutral, the valine will be negatively charged, and the arginine will be positively charged.

845. C At a pH of 5.4 the asparagine is neutral, the aspartic acid is negatively charged and the glycine is positively charged.

846. C The fact that asparagine's pI is lower indicates that its side chain is less basic; however the table shows that the side chain does contain a nitrogen, eliminating answers A and B. The difference lies in the fact that nitrogen atoms of amides are less basic than nitrogen atoms of amines.

847. B The side chain (circled below) matches valine's side chain.

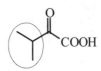

848. D The side chain (circled below) matches asparagine's side chain.

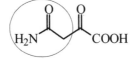

849. C According to the table, tyrosine contains a phenyl group and answer C is the only ketone that has a side chain with a phenyl group.

850. C The hydrogens can be added to the imine from either face producing a 50:50 mixture of R and S enantiomers.

851. C The amino acids found in the body are all α-amino acids, so the ketone must be on the carbon next to the carboxylic acid to produce an α-amino acid.

852. B The pI must be between the two pK_a values and only answer B falls between these two values.

853. B The proton of the carboxylic acid is more acidic than the proton of the protonated amino group.

854. A Only answer A has the formula $C_n(H_2O)_n$.

855. C Fatty acids are stored as triglycerides.

856. A Glucose is an aldehyde.

857. C Glucose contains 6 carbons, so 6 moles of carbon dioxide would be produced.

858. C Fructose contains 6 carbons, so 6 moles of carbon dioxide would be produced.

859. C Only D-fructose is assimilated into the body.

860. B Glucose has the pattern with the last two hydroxyl groups on the same side (circled below). In L-glucose, they point to the left.

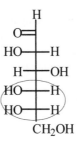

861. D Fructose has the pattern with the last two hydroxyl groups on the same side (circled below). In L-fructose, they point to the left.

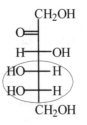

862. A Fructose has the pattern with the last two hydroxyl groups on the same side (circled below). In D-fructose, they point to the right.

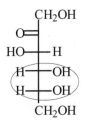

863. C D and L glucose are non-superimposable mirror images of each other.

864. D Although L-fructose comes from L-glyceraldehyde which does rotate light counterclockwise, L-fructose does not necessarily rotate light counterclockwise.

865. B There is a hydroxyl group and an ether attached to the same carbon forming a hemiacetal.

866. D Answer D is the ring with the OH groups on the proper side of the ring.

867. A The carbonyl carbon or the hemiacetal carbon in the ring is considered the anomeric carbon.

868. C The hydroxide of the fifth carbon attacks the carbonyl to form the pyranose ring.

869. C The two structures are only different at the anomeric carbon.

870. A The two are non superimposable mirror images.

871. A Only the structure in answer A has the hydroxyl group on the highest number chiral carbon pointing to the right.

872. A The two structures are diastereomers that differ at only one carbon.

873. C Both have the hydroxyl group of the highest chiral carbon facing to the right.

874. C Both are found in nature because they are D sugars.

875. C The priorities are marked below. Both carbons rotates counterclockwise, however, the hydrogens are coming out of the page.

876. B Threose has the opposite configuration at carbon 2 and the same configuration at carbon 3 as erythrose.

877. C Threose is an aldehyde (I) with 3 carbons (III).

878. A The last two hydroxyl groups on glucose point in the same direction which matches erythrose.

879. A In the statement above the sugars, it states that D-glyceraldehyde is (+) which indicates that it rotates plane polarized light clockwise.

880. B The direction of light rotation (III) cannot be determined without measurement.

881. B There are 3 chiral carbons so there are 2^3 possible stereoisomers.

882. A Ribose differs from arabinose at the second carbon only.

883. C Xylose and ribose differ only at the third carbon.

884. A The hemiacetal carbon is carbon 1.

885. A The opening of the ring is shown below.

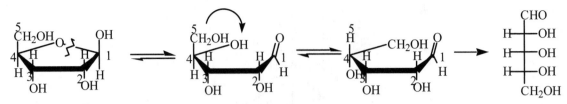

886. C There is a hemiketal carbon instead of a hemiacetal indicating that the ring came from a ketose.

887. D D-fructose is formed by shifting the carbonyl of glucose by one carbon. The shift does not affect the highest numbered chiral carbon, so D-glucose would form D-fructose.

888. A Aldehydes reduce to alcohols.

889. B The product has a plane of symmetry through the center.

890. C The hydrogen can add to either face producing both R and S configuration.

891. A D-glucose is the only aldehyde listed.

892. C Sucrose is composed of glucose and fructose. If both rings are opened as in the answer to question 885, the structures can be seen.

893. A The pyranose (6 membered ring) is on the top and the linkage is α-glycosidic because the oxygen atom connected to the anomeric carbon is facing down.

894. B The furanose (5 membered ring) is on the bottom and the linkage has the oxygen atom attached to the anomeric carbon is facing up, so it is a β-glycosidic linkage.

895. C Because the acetal and ketal are less reactive to hydrolysis, the chain does not open producing an aldehyde with which the Tollens reagent can react.

896. A Cellulose is made up of sugars making it a carbohydrate.

897. A The linkage at the anomeric (acetal) carbon is up making it an α linkage. The linkage that is down is not at an anomeric carbon.

898. A The carbon is attached to two ethers and a hydrogen.

899. D The R replaces the OH group which indicates that the furanose (five membered ring) is of an aldose because an H is attached to the anomeric carbon.

900. B See the ring opening in answer 885.

901. D The products are diastereomers that are different at only one carbon.

902. B A carboxylic acid is at a higher oxidation state than an aldehyde.

903. C The direction cannot be determined without being measured.

904. D Meso products do not rotate plane polarized light.

905. A The sugar has the hydroxyl group on the highest numbered chiral carbon to the right so it is a D sugar. It is an aldehyde, so it is an aldose.

906. B If the nonchiral hydroxyl group on the end of the chain attacks the carbonyl a five membered ring will form. A six membered ring cannot form.

907. A If the linkage is at carbon 4 then the anomeric carbon is at carbon 1.

908. A The linkage is at the anomeric carbon of galactopyranose and it is oriented up. Therefore it is a β-linkage.

909. C See the ring opening procedure in answer 885.

910. D The two structures vary only at the fourth carbon.

911. B The units of chemical shift is ppm (parts per million).

912. D NMR stands for nuclear magnetic resonance.

913. B In distillation the mixture is boiled and each component vapor is collected.

914. A In chromatography, a silica gel column or plate is used to separate compounds based on their affinity for the silica gel.

915. D Extraction uses two layers (organic and aqueous) to separate compounds by solubility.

916. A The name IR comes from infrared.

917. A The R_f value is calculated taking the distance traveled by the compound and dividing by the distance traveled by the solvent.

918. A Chemical shifts in NMR spectrum result from different shielding and deshielding due to electronic environment.

919. B The splitting of NMR peaks results form the neighboring hydrogens. There are n + 1 peaks, where n is the number of neighboring hydrogens.

920. C The units of IR peaks are wavenumbers (cm^{-1}).

921. A R_f values must be between 0 and 1 and answer A is the only value between these two numbers.

922. D An R_f of zero indicates that the compound did not move. Only very polar compounds do not move on silica gel. See answer to question 917.

923. **A** An R_f value of 1 indicates that the compound moved with the solvent, therefore it must not be very polar. See answer to question 917.

924. **C** Integration of the peaks (area under the peaks) gives the relative number of protons contributing to the signal.

925. **B** The stiffer a bond, the higher the frequency of vibration.

926. **B** NMR peaks result from the absorption of energy in the radio frequency range of the electromagnetic spectrum.

927. **B** There are five different types of hydrogens which are circled on the structure below.

928. **B** There is only one peak with no neighboring hydrogens and it is circled in gray above.

929. **B** IR spectrum deal with the vibrations of bonds.

930. **A** The spin of the nucleus on the hydrogen atoms flip to produce an NMR spectrum.

931. **B** The O-H stretch happens at 3400 cm^{-1}.

932. **A** The C = O stretch happens at 1700 cm^{-1}, and a ketone is the only functional group listed with a carbonyl present.

933. **A** There are five types of hydrogens in 2-hexanone. Carbon 2 does not have any hydrogens.

934. **D** There are 3 hydrogens on carbon 1, so the peak will integrate to 3. There are no neighboring hydrogens so the peak is a singlet.

935. **B** There are 2 hydrogens on carbon number 3, so the peak will integrate to 2. There are two neighboring hydrogens, so by the n +1 rule the peak will be a triplet.

936. **D** The carbons farthest upfield are the hydrogens farthest from the carbonyl.

937. **A** The carbonyl will produce a peak at 1680 cm^{-1}.

938. **D** Propyl alcohol has the lowest boiling point, so it will be the first to come across in a distillation.

939. **A** Benzyl alcohol has the highest boiling point, so it will be the last to come across in distillation.

940. **A** Benzyl and propyl alcohols have the largest difference in boiling point, therefore they are the easiest to separate using distillation.

941. **C** Benzyl and octyl have the closest boiling points, so they will be the hardest to separate.

942. **B** The two have boiling points that are only different by 1°C so they cannot be separated by distillation using distillation.

943. **A** The ester and the alcohol can be separated because the boiling points are 20°C different.

944. **B** There is a plane of symmetry in morpholine making only three different types of hydrogens circled below.

945. **B** Both of the peaks from the hydrogens connected to the carbons will be triplets because each peak has 2 hydrogens on the neighboring carbons.

946. **D** The N-H stretch occurs at a high frequency.

947. **C** The amine will be in the aqueous layer after the first step because the strong acid (HCl) will make the base (the amine) soluble in the aqueous solution.

948. **B** The amine is protonated which makes it charged and more water soluble.

949. **D** The benzene is the only component that is not removed by any step of the separation because it cannot be protonated or deprotonated.

950. **C** The benzoic acid is deprotonated by the weak base, but the phenol is not. The charged anion is more soluble in water.

951. **D** The exchange of steps 1 and 2 will not change the separation because one component is still removed each time.

952. **A** If steps 2 and 3 are reversed, the strong base will deprotonate both the phenol and the benzoic acid removing them together in the new step 2.

953. **B** Peak D is a peak from TMS a reference put into NMR samples. The purpose is to reference the chemical shifts to 0 ppm.

954. **B** Peak A is a triplet, so it must have 2 hydrogens on the neighboring carbon. Splitting follows the n + 1 rule where n is the number of hydrogens on the neighboring carbon.

955. **A** To the left in a NMR spectrum is downfield.

956. **B** The peak down at 9-10 ppm indicates the hydrogen of an aldehyde. It is an unusual peak that far downfield and it should be memorized.

957. **D** Peak D is the farthest to the right, therefore, it is the farthest upfield.

958. **A** The farther to the left a peak is the more deshielding of electrons a proton "feels".

959. **B** Peak B is split into the most peaks, so it has the most neighboring hydrogens.

960. B Peak B is split into four sets of two (quartet of doublets) indicating two different types of protons responsible for the splitting.

961. D Reading from the axis of ppm the peak is at 9.8 ppm. δ stands for chemical shift.

962. B There are three peaks in addition to peak D, the reference peak, indicating that there are three different "types" of hydrogens.

963. A The hint that there is an aldhyde helps to select A. However, only the structure in answer A corresponds to the spectrum. The different peaks are shown below.

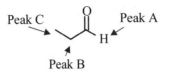

964. A The most polar compound is the one that has traveled the shortest distance.

965. D The least polar compound is the one that has traveled the farthest.

966. D The compound that has traveled the farthest will have the greatest R_f value.

967. A The compound that has traveled the shortest distance will have the smallest R_f value.

968. A The R_f value is calculated by dividing the distance from the starting point to the spot of the compound by the distance from the starting to the solvent front.

$$R_f = \frac{80\,mm}{100\,mm}$$

969. A There are five peaks in addition to the reference peak (peaks A – E) so there are five different "types" of hydrogens on the compound.

970. A The peak is to the right of 1.0 ppm, therefore, it must be 0.9 ppm.

971. A Peak A is the farthest downfield because it is the farthest to the right.

972. B Peak C is split into the most peaks, so it must have the most non-chemically equivalent neighboring hydrogens.

973. D The numbers above the peaks corresponds to the integration. Peak E corresponds to the most hydrogens because it has the greatest integration.

974. C Peak A is split into three peaks, therefore, it is a triplet.

975. B Peak A integrates as 2, so 2 hydrogens produce the signal.

976. B Since peak A is a triplet, there must be 2 hydrogens on the neighboring carbons. Splitting follows the n+1 rule where n is the number of neighboring hydrogens.

977. A The integration is the area under the peak.

978. A Peak A is the farthest downfield, therefore it has the most deshielding which indicates proximity to electronegative atoms.

979. A The hydrogens on carbon 3 have two neighboring hydrogens, therefore, the peak will be a triplet. Peak A is the only triplet. The carbon is attached to an oxygen which shifts the peak downfield.

980. C Peak C integrates as one hydrogen and carbon 5 is the only carbon with one hydrogen. Peak C also has many neighboring hydrogens causing it to be highly split.

981. C Peak E integrates to six hydrogens and is a doublet (one neighboring hydrogen). Carbon 6 and carbon 7 are chemically equivalent due to symmetry.

982. A The caffeine has traveled the shortest distance indicating that it is the most polar compound.

983. C Acetaminophen traveled the farthest, so it will have the greatest R_f value. See answer 968.

984. B The spot for A has not traveled more than halfway, yet it has moved from the original line, so it must have an R_f between 0 and 0.5. Answer B is the only value that falls into that range.

985. D Since the Excedrin lane has three spots matching the out spots in lanes A, B, and C, it contains all three substances.

986. B The Tylenol lane only has a spot matching the acetaminophen spot.

987. A By increasing the amount of the polar solvent, all R_f values are increased because the polar substances will travel more readily in a more polar solvent.

988. A Ethanol that is purified by distillation will be 96% ethanol because the mixture being distilled becomes an azeotrope. An azeotrope is a mixture which boils at a lower point than either pure compound.

989. C The goal is to create a saturated hot solution. With a solubility of 1 g/20 ml, 5 grams will need to be dissolved in 100 ml.

990. B Recrystallization never recovers all of the starting material because it has some solubility in the solvent at room temperature.

991. A The salt $MgSO_4$ is an ionic compound that will be soluble in water.

992. C Both biphenyl and triphenyl carbinol are soluble in an organic solvent.

993. A If the density of the organic solvent is less than 1, it will sit on top of the water.

994. C The desired product is not soluble in hexane and both products are soluble in ether. The correct ratio of these two solvents would be the best choice for recrystallization.

995. A The sodium hydroxide removes a proton from the caffeine making it neutral and it also removes a proton from the phenols (tannins) and carboxylic acids making them charged. Only the neutral caffeine will be soluble in the organic layer.

996. C See the answer to 995.

997. A Without the sodium hydroxide, the caffeine will remain charged and the other two components neutral making only the caffeine soluble in the aqueous phase.

998. C See the answer to 997.

999. C Since the bicarbonate is a weak base, it would not deprotonate the phenol, but it would still deprotonate the caffeine. Thus, both the caffeine and tannins would be neutral and therefore soluble in the organic phase.

1000. B See the answer to 999.

1001. B Organic solvents such as dichloromethane with a density greater than 1 will be on the bottom of an extraction.